A CASE FOR WETLAND RESTORATION

DONALD L. HEY
The Wetlands Initiative

NANCY S. PHILIPPI
Philippi Associates

A Wiley-Interscience Publication
JOHN WILEY & SONS, INC.
New York / Chichester / Weinheim / Brisbane / Singapore / Toronto

This book is printed on acid-free paper. ∞

Published simultaneously in Canada.

For ordering and customer service, call 1-800-CALL-WILEY.

Library of Congress Cataloging-in-Publication Data

Hey, Donald L.
A case for wetland restoration / Donald L. Hey and Nancy S. Philippi.
p. cm.
Includes bibliographical references.
ISBN 0-471-17642-7 (alk. paper)
1. Wetland conservation. 2. Wetlands. 3. Restoration ecology.
4. Wetland conservation—United States—Case studies. I. Philippi, Nancy S. II. Title.
QH75.H49 1999
333.91′8153′0973—dc21 98-53674
CIP

Printed in the United States of America

10 9 8 7 6 5 4 3 2 1

CONTENTS

PREFACE

We were struck, as we began this study, with the negative press that wetland restoration was getting among its practitioners as well as the general public. Creation and restoration of wetlands is a new science, and there is no question that we have a lot to learn. The more important story, we now are convinced, is how much we have already accomplished. After only two decades of experimentation, the field has exploded—the number of restoration projects has increased, the body of knowledge has expanded, and the methodology has improved in ways that promise the imminent achievement of our national goal of no net loss and rapid progress toward our long-term goal of increases in the quantity and quality of our wetland resources.

The federal programs that have been adopted since the early 1970s have played a major role in making this happen. Although the U.S. Army Corps of Engineers' implementation of the Section 404 program is much maligned in the environmental community, the program has had a profound effect on the reduction in our wetland losses and the mitigation of what losses remain with solid and creative restoration projects. The record is a good one and it improves daily.

Even the failures are successes of a sort. Land set aside for wetland restoration is permanently protected from development, and its potential for providing wetland benefits remains intact. If the technology was inadequate to fulfill that potential in the past, it will no doubt be able to meet the task at some point in the future. The preserved landscape will still be there.

Our experience with the projects described in this book has brought us to this optimistic conclusion. Considering the problems they encountered, the rudimentary state of restoration technology at the time they were implemented, and the disorganized and inexperienced bureaucracy of federal programs that guided their development, it is surprising that they accomplished anything at all. On the contrary, each of them accomplished a great deal.

The selection of the four case studies deserves some explanation because the problems encountered in making those choices reflect some aspects of the art of wetland restoration's current predicament in answering the question: Is wetland restoration successful? The authors contacted several dozen well-known practitioners in the field in search of large restoration projects in different parts of the country that had been judged by their implementers to be successful in meeting their original design goals. With the exception of a few respondents who apparently thought the question to be frivolous, and a few others who said nothing works, public agency staff and private consultants made every effort to provide us with suggestions. The most typical response to our request was for them to predict that there were many projects that met our criteria, only to come up with one or two or, in too many cases, none that actually did. The mitigation projects that had been in the ground long enough to provide a clear record of results were, in most cases, planned before the process demanded clear-cut goals and careful monitoring for success, as they usually do today. In another 20 years, however, an exercise such as this should be exceedingly fruitful.

The Wetland Reserve Program, for example, is only a few years old and their projects are just now being put in the ground. Partners for Wildlife projects, on the other hand, have been going on for years but, although the field practitioners have a sense of great success, there has been relatively little monitoring or consideration of either quantitative or qualitative goals. As one field worker told us, "we never set goals because that makes it too easy to be judged a failure. The fact is, we get great projects anyway."

We visited and reviewed six projects nationwide before we settled on the four discussed here, largely for the geographical range they represent. Their uniqueness impressed us and, had we included the other two, we would have had even more diversity of wetland type and mitigation issues.

We ended the study, therefore, with far more positive feelings about the success of wetland restoration than we had had at the beginning. As frustrating as the Section 404 and NEPA processes may be to those who grapple with them, they seem to work. People battling out the tradeoffs between separate and conflicting interests is, after all, our democratic way. The better, the smarter, and the more committed those people are, the more successful that way is.

We are grateful to the Federal Highway Administration for providing funding for our research, and to their project manager, Dr. Paul Garrett, who helped to select the case studies and reviewed the results. We owe special thanks, as well, to Bonnie Kath for her patience and expert word processing and to Spencer Hey and Kevin Kleinjan for their excellent computer graphics skills.

DONALD L. HEY
NANCY S. PHILIPPI

The Wetlands Initiative

A CASE FOR WETLAND RESTORATION

CHAPTER 1

INTRODUCTION

Wetlands can be restored, of that there is no doubt. They can be restored to provide the functions that have been lost—of that too, there is no doubt. A case is easily made that defends these statements in the following chapters. But are they always restored when the attempt is made and do they always replace the very functions that are lost? Of course not. More times yes than no, or no than yes? It would be even more fruitless than it would be arrogant to hazard such a guess.

It is far too early in the development of wetland restoration, as a science, to make pronouncements on its overall success or failure. The reader will not find here a comprehensive statistical proof that restoration is all that it can or should be but, rather, a description of what the country needs from wetland restoration, an explanation of how and why wetland restoration has become important, and an examination of four successful wetland restoration projects for what they might tell us about the ingredients that contribute to success.

A single question, central to the discussion, remains to be answered much further in the future: Are projects that restore wetlands to compensate for wetland losses successful? Yet this question cannot be answered without responding to a second one: What makes a restoration project successful? This question, in turn, must be broken down into two more: What elements have to be present in a restoration project for it to be perceived as successful, and What conditions contribute to those elements?

An examination of the four restoration projects, one of them initiated in the 1960s, a second in the 1970s, and all four of them completed in the 1980s, will provide some answers to these last two questions. It will even shed light on the question of "what makes wetland restoration successful?" but it will not provide the final answer. Likewise, it will add to the body of knowledge and opinion that will generate, sometime in the future, an answer to the question of whether restored wetlands truly compensate for the losses of natural wetlands.

The first thing that is examined in the following pages is the route by which the United States arrived at the place in history where these questions are even important. That route began in the natural landscape of North America, before European settlers had arrived, when close to 392 million acres, or more than 11% of the total continental land mass, was wetlands (Dahl, 1990). The economic development that drove the nation's history through the eighteenth and nineteenth centuries caused large acreages to be drained and tiled for agricultural use, and the navigational and, later, flood control needs of the expanding new nation were met by dredging, channelizing, and leveeing the nation's rivers. It was these processes by which about 53% of the original wetlands were drained and their functions lost.

The functions of wetlands were neither appreciated nor even understood as early Americans drained and filled the original wetlands. It was only after sportsmen and wildlife enthusiasts noticed the diminishing numbers of migratory waterfowl traversing the flyways that they connected this decrease with lost wetland habitat. The agricultural community eventually became concerned about the high-energy streamflows and heavy floods that eroded their fields and washed away the topsoil, but only after the tile drainage had been put in place that contributed to the damage. Urban dwellers noticed that once-abundant water supplies had begun to dry up only after those holding reservoirs, the wetlands, had themselves dried up and been filled in. And water resources planners asked why flood damages were increasing along the nation's major rivers, no matter how many flood control works had been installed, realizing too late that the wetlands that once captured and then slowly released the heavy flood flows had been cut off from river channels by navigation and flood control projects. Finally, by the time the nation moved to curb the pollution of the surface waters, in the environmental decade of the 1970s, more than half of the wetlands that had once filtered out so many sediments, nutrients, and other pollutants had been destroyed.

As the twentieth century progressed, each of these realizations prompted a federal action. The Fish and Wildlife Service (FWS) was given congressional authority to purchase wildlife refuges that would provide habitat for migratory waterfowl. The U.S. Department of Agriculture (USDA) stopped funding drainage projects and even, quite recently, withheld federal subsidies from farmers who drained new lands. Most importantly, the U.S. Army Corps of Engineers (the Corps) was authorized to require permits from developers who intended to damage or destroy wetlands. This permit process, under Section 404 of what was eventually to be known as the Clean Water Act, gave the U.S. Environmental Protection Agency (USEPA) a major role in requiring the permit applicants to avoid, to minimize, or at the very least, to compensate for any wetlands they destroyed to pursue their economic development ends. All these federal actions—lodged in four major federal agencies—constitute a collective body of law that attempts to protect and preserve the chemical, biological, and hydrologic functions of the wetlands that remain today.

This national policy was not arrived at by, nor does it sit easily with, all segments of the population. Farmers, developers, and others have protested the prohibition upon the use of land to which they hold legal title. Under the fifth amendment of the Bill of Rights, the property owner must be compensated for any of his land "taken" for public use. Prohibiting the draining of wetlands, the argument goes, is the equivalent of taking it for public use.

The concern about wetland losses in the United States peaked with a National Wetlands Forum in 1987. The forum established a national wetlands goal that has been incorporated, subsequently, into all major national wetland policy and legislation: to achieve no net loss of the nation's remaining wetlands base and to restore and create wetlands, where feasible, to increase the quality and quantity of the nation's wetlands resource base.

At the time this goal was approved, the nation was losing an estimated 290,000 acres of wetlands annually. This was a vast improvement over the 20-year period immediately preceding the implementation of the Section 404 permitting process, during which almost 500,000 acres per year were being destroyed. An update in the 1990s suggests that the annual loss of wetlands has been reduced even further, to approximately 100,000 acres annually.

The challenge at the beginning of the 1990s, therefore, was to transform that continuing loss of 100,000 acres into "no net loss," and the solution was to create new wetlands and to restore those wetlands that had been previously drained. Wetland restoration, as a science, has gained a lot of ground since the 1970s and is being used extensively today to compensate for (mitigate) the wetlands that are being lost.

No one had ever intended that the Section 404 process would totally prevent the loss of wetlands—only that it would require the avoidance or minimization of losses wherever possible. There are, were, and will continue to be development activities that require water-related locations and that are sufficiently important to justify the issuance of a permit. Where these activities result in a loss of wetlands, the permitting authority, the Corps, requires that compensatory wetlands be created either on site or elsewhere to replace the functions lost by the permitted action. The Corps' mitigation policy generally requires that more acres be restored than are being lost, by a ratio of at least 1.5:1 and sometimes greater. And a relatively new concept, the mitigation bank, is being promoted as a mechanism by which the permit applicant can simply "buy credits" in an off-site, established, and preapproved bank. Wetland restoration is used by other agencies—by the FWS in their Partners for Wildlife, for example, and by the USDA in the Wetland Reserve Program (WRP). Both these programs depend upon the voluntary restoration of wetlands by landowners, encouraged by monetary incentives provided by the federal agencies.

Restoration of former wetlands, therefore, either as compensation for wetlands destroyed by Section 404 permit holders or as voluntary actions of landowners participating in the Partners for Wildlife and WRP programs, will furnish the mechanism by which the national wetland goals can be achieved. The only problem remaining, in the late 1990s, is that some observers and even practitioners of the restoration science are skeptical about the ability of restoration projects to replicate truly the functions of natural wetlands. Horror stories of poorly designed projects, failed projects, and promised but never-executed projects are common in the literature. The hydrologic, chemical, and biological functions of wetlands have not been replaced by restoration projects, critics claim, because of bad or inadequate science and faulty or sloppy execution.

This is where we are today—asking if wetland mitigation is really compensating for the loss of natural wetlands—and the answers that come in are yes, no, and maybe. Because so few restoration projects have been in the ground long enough to show results, the most productive source of answers to that question seemed to lie in taking a careful look at a few of these older projects, projects that have been perceived by their sponsors as "successful," and to ask why they were being called successful and how they got that way. This exercise constitutes the latter half of this book.

The four cases that are described here (Fig. 1-1) are as follows:

- The mitigation of damages incurred by the replacement of 37 old and unsafe bridges on the Florida Keys Overseas Highway that stretches from the tip of the Florida peninsula to the southwestern-most island of Key West. Major damages anticipated from the action were the destruction of seagrass and mangrove communities, increased potential for shoreline erosion, and interference with hydrologic exchanges between the Florida Gulf to the north of the Keys and the Atlantic Ocean to the south.
- The mitigation of damages incurred by the construction of a six-lane limited access roadway, the Beltline, to skirt the south side of Madison, Wisconsin, to alleviate traffic congestion created by the increase in traffic from 5,000 to 50,000 trips per day. The highway was designed to transect the Yahara River Marshes and, by destroying portions of them, would affect some of the area's most valuable wildlife habitat.
- The mitigation of damages incurred by the widening and straightening of a stretch of mountain highway west of Denver, Colorado, to improve the safety of a road heavily used by traffic headed for the nation's major ski resorts. The improvements required the straightening of the North Fork Platte River and the loss of alpine riparian wetlands.
- The mitigation of damages anticipated from the construction of a 280-acre commercial development in Yolo County on the Sacramento River in California. The development was designed to bring much needed jobs, economic development, and tax revenues to the fast growing region. It required the destruction of riparian habitat along the river, of wetlands, and of the habitat of the endangered elderberry valley beetle.

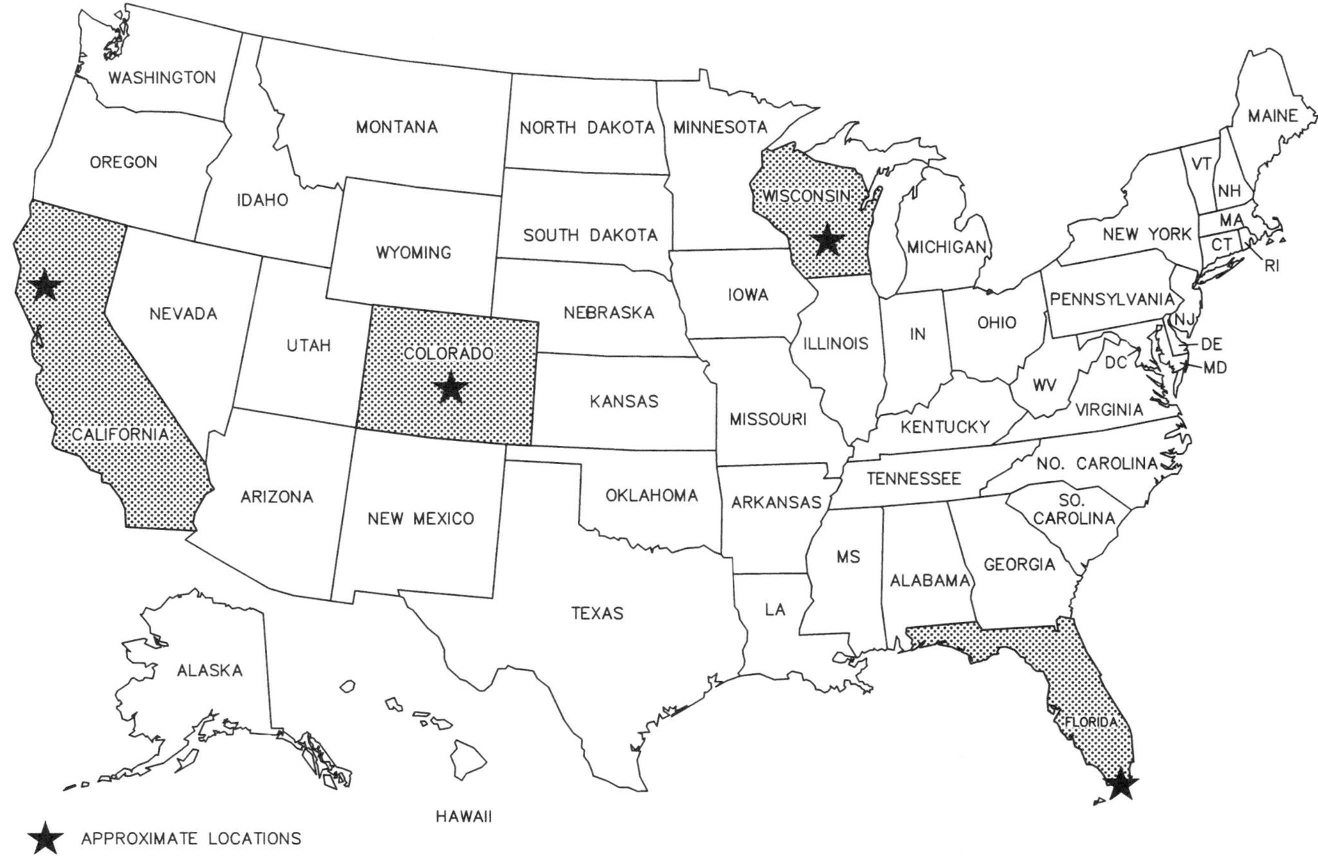

Figure 1-1 Project distribution.

All four projects were perceived as successful by those that had participated in their implementation. In each case, the environmental review and negotiation process set in motion by the National Environmental Policy Act (NEPA) and Section 404 proved successful in preventing environmental losses. Although quantitative assessments were relied upon far more than qualitative assessments, for both the affected and the newly created wetlands, a net gain of wetland functions characterized all four cases. The state of the art (or science, depending on the point of view) prevented definitive answers to the questions of whether the functions gained were equivalent to those that were lost, or even whether they should have been. With the rapid escalation of our scientific understanding of wetland processes and restoration techniques, along with the body of knowledge being developed in our public and private institutions, the answers to those questions will no doubt be available in another 10 or so years, as the results start coming in from projects being started now. In the meantime, the answers coming from the four case histories indicate great promise for the future of wetland restoration in the United States.

REFERENCE

Dahl, T. E., *Wetlands Losses in the United States 1780s to 1980s*, U.S. Department of the Interior, Fish and Wildlife Service, Washington, DC 1990.

CHAPTER 2

WETLANDS IN THE NATURAL LANDSCAPE

Since the last ice age, wetlands, to one degree or another, have existed in every state, from Maine to Florida to Hawaii. Wetlands are even found in the more arid west, including Utah, Nevada, New Mexico, and Arizona, and in the frigid climes of Alaska. However, wetlands are more abundant in regions where precipitation exceeds evapotranspiration and where the topography is reasonably flat with soils underlain by impermeable materials such as clay or limestone. Wetlands readily form where groundwater is high or discharges to the surface, as with peatlands and fens. Consequently, wetlands can be found on hillsides, in glacial valleys, on floodplains and fluvial fans, and along shorelines.

The formation of a wetland is a reasonably simple process. It involves retention of water above, at, or near the ground surface for periods of time long enough to cause anoxic (oxygen lacking) conditions to form at and below the soil–water interface. Given modest variations in these conditions, a wide variety of wetland types can form. For example, bogs require a confined draining system, deep depressions, and continuous saturation. A sedge meadow requires a little less water: shallow inundation and seasonal dry periods. The establishment of these various conditions is due to a number of factors, some better understood and more widely accepted than others.

Along our coastal zones (excluding the coastline of the Great Lakes), topography and tidal cycles define the relationship between soil and water. Further inland, wetland conditions are defined by the interaction of landforms with freshwater and tidal influences. The principal driving function of interior, freshwater wetlands is hydrology. Whether palustrine, lacustrine, or riverine, freshwater wetlands are driven by the rainfall–runoff–groundwater relationships for their geographic location. In the pothole region, spanning the glaciated Midwest from Iowa to Alberta, Canada (Van der Valk, 1989), although wetlands could not occur without the surface depressions of this complex topography, the ambient hydrology provides the sufficient, defining condition. Similarly, the presence of a lake establishes the necessary conditions, but the level of the lake's water relative to its shoreline determines the nature and extent of wetland fringe.

A number of factors operate in and on the landscape to form the appropriate hydrologic conditions for wetlands. These factors affect the movement and distribution of water in space and time. They are both geomorphic and biotic. The fidelity with which these factors are copied determines the path of restoration and the degree of success.

DEFINING FACTORS

Governmental regulations as well as the physical environment define wetland factors. The regulatory definition involves hydrology, soils, and plants, but these three components are not independent. Wetlands originate at appropriate intersections among topography, hydrology, and geology. And the particular soils and plants, in large part, result from the in situ geology and ambient hydrology.

Many landforms impound or control water to sufficient extent that wetlands form as a result. The depressions left behind by the retreating ice sheet of the Wisconsin glacier serve as ideal receptacles for retaining water and nurturing the formation of hydric soils and the propagation of hydrophytes (water-loving plants). The natural levees bordering the Sacramento River trapped floodwaters in the overflow areas. These waters defined and supplied the vast associated marshes with their critical nutrients. Glacial moraines, eskers, potholes, and lakes, along with fluvial features, such as swales, streams, rivers, bars, natural levees, and fans, and marine landscapes such as intertidal mud flats, tidal pools, aquatic beds, and coral reefs all form the infrastructure of North American wetlands. The inhabiting plants and animals, then, modify these structures.

From a structural point of view, the species of plants have little effect on the formation and survival of wetlands. Forming on the in situ geologic and topographical base, and constrained by the hydrologic regime, it is the root structure, stem density, and plant mass that are the important factors. The effects of these factors extend to both living and dead plant materials. Living plants in streams, lakes, and estuaries slow the movement of water, increasing water depths and the area of inundation. Plant detritus has a similar effect. It displaces water and increases boundary friction. Plants affect the hydrologic cycle, the degree to which depends on the community type (e.g., grasses vis-à-vis trees). Interception storage and transpiration are both affected by plant physiology. In the soil profile, organic material retains water in the vadic (unsaturated) zone, critical to the propagation and survival of hydrophytes. Plants and plant detritus provide the expansive stable surfaces and niches supporting the macro- and microorganisms responsible for decomposing and recycling nutrients and other chemical constituents in the water and providing habitats at the base of the food web. Through their control of water depth, and thereby anoxic conditions, and their contribution of organic matter, plants are important factors in the formation of hydric soils.

The significance of the "plant factor" should not be underestimated. One illustration of the importance of plants is peat deposits. Moreover, there are numerous historical references to the intensive and extensive drainage controls due to fallen trees and other plant debris. In 1850, the French geographer Desor (1879) explored the Upper Peninsula of Michigan. He encountered numerous debris dams as he traveled up the Monistique River. He reported that

> Such barriers are not rare in the forest, particularly if the inclination of the river is small. A trunk being carried by the river suddenly can be caught in the middle of a meander. If the stream is not strong enough to move it, it stays there and a second trunk comes to attach itself to the first one and many others come along and finally their branches get intertwined and they finally form a dam which can get bigger and bigger ad infinitum. Some of the barriers are rather big and seem to be rather old because you can find trees growing on top of the floating trunks.

He reported broad swampy areas all along the course of the river, undoubtedly due to the hydraulic controls of the debris dams. The structures viewed by Desor are dwarfed by one on the Red River in Oklahoma (Foreman, 1937):

> The phenomenon known as the Great Raft was a succession of log masses that choked the Red River for a distance of more than a hundred miles and was of unknown antiquity. It had existed so long as to assume permanent form and it was said that forest trees were to be seen growing upon it; horsemen could ride over it not knowing that they were passing over the water of the river. When removal of the Choctaw Indians was commenced in 1832, orders were given to attempt the removal of the raft so that navigation of the Red River could be established and supplies for the emigrating Indians could be brought up the stream. . . . Destruction of the raft was carried on under the command of Capt. Henry M. Shreve with a force of 150 men and four snag-boats. It was five years before Shreve could report the completion of the work.

Less massive debris obstructions were common through North American drainageways, from the Connecticut to the Sacramento Rivers. Geographic names often reflected their presence [e.g., Embarras (the French word for obstruction) Portage in Alberta, Canada, and the Embarras River in Illinois]. George Washington, in the 1780s, was a proponent of removing sandbars and snags on the Ohio River to improve navigation (Frost and Mitsch, 1989). And the ever-popular writer Samuel Clemens often lamented, in story and in life, the hazards of navigating the Mississippi and Ohio Rivers. He chronicled the dangers and the details of the numerous obstructions on the rivers. He even named one of his storied characters, Tom Sawyer, after the American term, *sawyer*, meaning a tree swept into the river with one end stuck in the mud and the other bobbing up and down in the current. The various states and the federal government worked to remove some obstructions in the late 1700s and early 1800s. But in 1827, the first Rivers and Harbors Act empowered the federal government to move in earnest. By the 1870s, few obstructions remained. As a result, hydraulic profiles were lowered and wetland areas reduced.

Still, debris obstructions are phenomenona of the modern drainageway and, despite the environmental harm, a good deal of effort is expended on their removal. The consequences are both physical and biological. Wetlands upstream of a log jam are drained, and sediment is flushed downstream. Habitats, for both plants and animals, are lost. These consequences are illustrated by the removal of a log jam on Locust Creek, a tributary of the Grand River, which is tributary to the Missouri River in north-central Missouri. A large volume of sediment had been impounded by the log jam and the backwater supported an abundance of wildlife. When the logs were cleared, the habitat was destroyed and the sediment was flushed downstream to smother the herbaceous layer of an extant forested wetland (Fig. 2-1).

At the same time that navigation was being improved, agricultural development was occurring west of the Appalachian Mountains and drainage obstructions were being removed from the small tributary streams to the larger navigable rivers. With the removal of beaver dams, detritus, and plants lining the swales and streams, drainage and navigation were improved while water levels were lowered and stream velocities increased on a continental scale. The results were numerous and some catastrophic: 100 million acres of wetlands were lost, flood damage greatly increased, water quality precipitously declined, and wildlife populations plummeted. These events serve to illustrate the importance of the "plant factor."

A number of animals are important to the creation and survival of wetlands. Birds, for example, move seeds from one wetland to another. Muskrats shape and harvest the plant community; bacteria and other benthic organisms shred and mineralize the detritus. But the organism most responsible for creating and preserving the prehistoric wetlands in North America is the beaver—European humans, on the other hand, are distinguished for the opposite effect. Beavers very purposefully retain water on the land surface for their own welfare and safety. Their dams, traversing swales, streams, and rivers, force water to spread across the adjacent and upstream landscapes and, by design, maintain shallow water depths, ranging from 2 to 3 feet over the central portion of the impoundment to only a few inches on the perimeter. These are ideal depths for a wide variety of wetland types—deep and shallow marshes, sedge meadows, and wet prairies. In a similar manner, beavers have controlled, and in some cases still control, the outlets or overflow structures of lakes and potholes influencing the presence and extent of the associated lacustrine and palustrine wetlands. Beavers also construct channels by which to reach and convey building materials and food supplies safely (Mills, 1913). The channels, some 1000 feet or

(*a*)

(*b*)

Figure 2-1 The effects of log jams on Locust Creek in north central Missouri. (*a*) Log jam impounding water and creating habitat. (*b*) Channel section after removal of log jam. (*c*) Sediment released by removal of log jam smothering herbaceous layer in forested wetland. Photographs by Ken McCarty, Missouri Department of Natural Resources. Photos also appear in color insert.

(*c*)

more in length, extend the hydrologic effects well beyond the limits of the impounded water.

The significance of the "beaver factor" is not well established; however, some observations and population estimates shed a little light on the subject. In the 1930s, two geomorphologists, studying streams in the Adirondack Mountains, concluded that beavers were the geologic agent responsible for the creation of the region's drainage systems (Ruedemann and Schoonmaker, 1938). They theorized that the level floodplains, which were perpendicular to the stream channel but stepped longitudinally, were the artifacts of beaver dams. Beaver dams trapped eroded materials, building, in sequence, marshes, meadows, and ultimately, drier floodplains. Although there is not a systematic body of knowledge validating this theory, scattered evidence supports the notion (Butler, 1995).

In an essay by Charles Davis (1907) on the origin and distribution of peat in Michigan, several interesting observations were cited, one by the geographer previously cited, E. Desor (1879):

> . . . in Michigan, rivers of considerable size which are barred by dams, making thus a quantity of lakes and ponds which would not exist without them. It is evident from this that, without these dams, the lake and peat deposits, which are found at the bottom of these ponds, would be less numerous. The beavers have thus exercised an influence not only on the distribution of waters, and the consequent fertility of the soil, but also up to a certain point even upon the distribution of recent rock formations.

Davis quoted Bela Hubbard, a land surveyor, on his observations in 1888 in the Detroit area:

> Not one or two, but a series of such dams were constructed along each stream so that very extensive surfaces became thus covered permanently with the flood. The trees were killed and the land converted into a chain of ponds and marshes with intervening dry ridges. In time, by nature's recuperative process—the annual growth of grasses and aquatic plants—these filled with muck or peat, with occasional deposits of bog lime, and the ponds and swales became dry again.
>
> Illustrations of this beaver-made country are numerous enough in our immediate vicinity. In a semi-circle of 12 miles around Detroit, having the river for a base, and embracing about 100,000 acres, fully one-fifth part consists of marshy tracts or prairies, which had their origin in the work of the beaver. A little further west, nearly a whole township in Wayne County is of this character.

Davis went on to explain succinctly the sequence of events and the results:

> If such an area was covered at the time when the water level was raised by a growth of trees or shrubs, these would be killed, and some would be cut down by the beavers for food and construction purposes; they would shortly be overgrown and buried. The destruction of this taller growth would enable the marsh

> plants bordering the stream and the water plants growing in it to spread out, first upon the margins of the flooded area, then because the water is shallow, over the whole impounded surface. Such an area of shallow water would fill rapidly with vegetable debris, and as reedy and grassy types of vegetation obtained a foothold and became abundant the water area would be restricted, until perchance, the animals built their dam higher or abandoned it for a new place.
>
> The fact that the original course of the stream had been obstructed by even a weak dam would tend to cause accumulations of drift material upon the obstructed area in time of floods, and eventually this might form such a check to the drainage that the water level might be raised faster than the vegetation could build up the surface. In such a case the tree growth would be destroyed, and again the area would be covered by water and marsh vegetation, to go through the cycles as before.
>
> In draining the extensive bog at Capac [Michigan], near the outlet end, or the south end of the marsh, beaver dams were cut through in making the ditches which had in the uppermost [dam] 4 feet, a second 10 feet, and a third 12 feet of peat over the top, while the section of the bog in the ditches was formed of successive and superposed layers of vegetable debris showing that several times forests had been succeeded by grass or sedge marshes. In its final stage this deposit was a Cedar and Tamarack swamp, with practically no natural outlet, the water which fell upon it either draining by seepage to the Belle river or evaporation from the surface.

The range of the beaver today covers all 50 states (being introduced to Hawaii) and the 10 provinces of Canada, excluding the arctic north, western deserts, and the peninsula of Florida. The habitat range comprises approximately 3.8 billion acres. In the conterminous United States, the habitat range of the beaver is approximately 1.9 billion acres, covering 85% of the land area. The range could have been even larger in the distant past. A fossil dam was found on the Seward Peninsula in Alaska (McCulloch and Hopkins, 1966). Embedded in peaty silt, the dam measured 10 feet high, 100 feet wide at the base, and 300 feet long. The location of this dam is north of the current timberline and outside the range of beaver habitation today, indicating the presence of beaver during a warmer period. The temporal range of beaver activity is extensive as well. Gnawed wood was found at three locations along Roberts Creek in Iowa (Chumbley et al., 1990), where remnants of a dam, made entirely of spruce, were uncovered. The wood dated 10,700 years b.p. Recently, beaver-gnawed logs were found buried under 4 feet of depositional soil material on a floodplain of the Des Plaines River in Wadsworth, Illinois (Hajac and Walz, 1998). The logs were found near the Pleistocene/Holocene interface. Carbon dating placed the age of the recovered woody material at 8500 ± 70 years b.p.

Within these spatial and temporal limits, the extent of wetland creation and maintenance by beaver depends on their numbers. Over their habitat range prior to the onslaught of the fur trade, the population estimates vary from 60 million to 400 million (Butler, 1995)—there likely were more beaver in North America than there are people today. For the most part, the structures created by these presettlement populations over the past 10 thousand years have been obliterated, at least from the land's surface, but their extent can be imagined based on modern observations. In a remote region of northern Minnesota, Naiman and Johnston (1988) reported a mean density of four beaver dams per mile or one every 1,300 feet. They noted that the dams were more frequent on streams along the Gulf of St. Lawrence in Quebec. There the density of dams ranged from 14 to 26 per mile. The mean density was 17, or one dam every 310 feet along the stream. The authors did not speculate as to what the presettlement density might have been or, more importantly, what area of surface water the beaver dams might have controlled. Still, behind every dam was a wetland.

The area of beaver-impounded water prior to the fur trade is unknown at this juncture. Yet the knowledge would provide an important perspective on the wetland resources of North America: how they were formed and how they functioned. Although there has not been a rigorous investigation of this subject, a rough idea can be formed from estimates of the beaver population, colony size, and area of water controlled by each colony. These estimates, for example, were applied to the Mississippi River watershed above Thebes, Illinois. In this 456 million acre watershed, the standing crop of beaver was estimated to be 40 million prior to the fur trade. In turn, the beaver population could have impounded 51 million acres of water (Hey and Philippi, 1995), which would account for 11% of the watershed area. In contrast, the wetland area in the 1780s was estimated to be 45 million acres (Dahl, 1990), 10% of the watershed. Another estimator of presettlement wetland area or beaver ponds is the presence of hydric soil. As of the twentieth century, approximately 40 million acres of hydric soils were present in the upper Mississippi, which accounts for 9% of the watershed. These three numbers (i.e., area of beaver ponds, wetlands, and hydric soil) are remarkably close given the divergent estimating techniques. Therefore, it is safe to say that prior to the fur trade the extent of wetlands could have been greater than the estimated area for the 1780s. Further, beaver could have played a significant role in creating freshwater wetlands in the United States. But whatever their former extent, the ways by which wetlands were lost suggest the means for their restoration.

LANDSCAPE POSITION AND STRUCTURE

Landscape position and structural form, along with hydrology, define the type of wetland. Wetlands are found on landscapes ranging from alpine slopes to ocean coastlines. This range has been subdivided and each element categorized by Cowardin et al. (1979), as shown in Table 2-1. The first two systems, marine and estuarine, function more under the influence of geomorphology and hydrology (tides) but can be affected by plants and animals. The other three systems, riverine, lacustrine, and palustrine, are defined by geomorphic, hydrologic, and biotic factors. A wetland can shift from one system to the other depending on the presence or lack of hydraulic controls exerted by biotic factors. In fact, for the same reasons, they can shift from one landscape position to another.

A small beaver dam constructed across a narrow valley can result in shifting the wetland type from riverine to lacustrine, as can be observed in northern Minnesota. A lacustrine wetland can be altered by the failure of a controlling dam and turned back into a riverine wetland. Palustrine wetlands can change from open water systems to deep marshes, to sedge meadows, and finally, to mesic prairie through the process of erosion and deposition of soil material.

Regardless of the landscape position of the wetland, its habitat and human values relate to its physical structure and nature: water depth, surface area, wetted surface, and frequency of inundation or desiccation. These parameters define the plant community, which, in combination, form the habitat structure. The plant species making up the botanical component of the structure, in turn, depend on the quantity of water and frequency of inundation or saturation as well as the typical quality of the surface and groundwater. The hydrology, plants, and edaphic (soil) organisms form the supporting soils.

The landscape position affects, to some degree, the societal benefits derived from a wetland. Wetlands serve best for flood control if they are located near an area of potential damage (Ogawa and Male, 1983). Likewise, a wetland near an outfall of a storm sewer or wastewater treatment plant can provide greater water quality benefits for those structures than one at a distance downstream, and no benefits would be provided by a wetland upstream of the outfall. Wildlife usage depends on the adjoining resources meeting the total needs of an organism. Lacking nearby trees, or other structures for roosting, bats will tend not to forage in a wetland (French, 1998). Without a safe route of ingress, primarily through stream channels or swales, mammals or fish are not likely to inhabit an otherwise suitable wetland. Position relative to need and supporting resources is an important consideration in creating or restoring a wetland, and the value of the wetland is inextricably tied to its position.

Species diversity undoubtedly was supported in a grand way by wetlands in the natural landscape. Today, however, plant diversity is too often taken as the sole criterion of successful restoration, relegating such critical functions as flood control, groundwater recharge, and water quality management to positions of secondary importance, if they are considered at all. These latter functions, if given greater consideration in restoration, can help sustain the aquatic ecosystem far beyond wetland boundaries. The capacity of a wetland to store water and prevent it from moving too quickly through the watershed can provide for sustained base flow, water quality treatment, and biodiversity. For example, alpine wetlands can be used to hold back runoff during spring melt and reduce the erosive force of high flows as they cascade downstream. At the same time, these wetlands can provide homes for beaver and forage for elk. As they provide flood storage, water quality is benefited as well. Generally, the longer the water is retained within a wetland the greater the opportunity for recycling nutrients and removing unwanted contaminants. During the long, dry summers,

TABLE 2-1 Cowardin's Categories

System	Subsystem	Class	Subsystem	Class
Marine	Subtidal	Rock bottom	Intertidal	Aquatic bed
		Unconsolidated bottom		Reef
		Aquatic bed		Rocky shore
		Reef		Unconsolidated shore
Estuarine	Subtidal	Rock bottom	Intertidal	Aquatic bed
		Unconsolidated bottom		Reef
		Aquatic bed		Streambed
		Reef		Rocky shore
				Unconsolidated shore
				Emergent wetland
				Scrub–Shrub wetland
				Forested wetland
Riverine	Tidal	Rock bottom	Lower Perennial	Rock bottom
		Unconsolidated bottom		Unconsolidated bottom
		Aquatic bed		Aquatic bed
		Rocky shore		Rocky shore
		Unconsolidated shore		Unconsolidated shore
		Emergent wetland		Emergent wetland

System	Subsystem	Class	Subsystem	Class
	Upper Perennial	Rock bottom	Intermittent	Streambed
		Unconsolidated bottom		
		Aquatic bed		
		Rocky shore		
		Unconsolidated shore		
Lacustrine	Limnetic	Rock bottom	Littoral	Rock bottom
		Unconsolidated bottom		Unconsolidated bottom
		Aquatic bed		Aquatic bed
				Rocky shore
				Unconsolidated shore
				Emergent wetland
Palustrine		Rock bottom		
		Unconsolidated bottom		
		Aquatic bed		
		Unconsolidated shore		
		Moss–Lichen wetland		
		Emergent wetland		
		Scrub–Shrub wetland		
		Forested wetland		

Source: Cowardin, 1979.

water can be released from subsurface storage to provide for the in-stream needs of fish, macroinvertebrates, and other organisms.

Although Dahl (1990) and others have made an attempt to quantify the number of remaining wetlands on a national scale, this information is not always available on a local scale, where it would be useful in mitigating for wetland losses. Knowing where and what type of wetland to create would greatly facilitate the overall ecological success of this nation's restoration activities. The necessary planning activities, such as are being pioneered in the state of Washington, must be improved and expanded. In the absence of this planning structure, individual restoration projects would benefit from an understanding of the landscape position and structure of the intended prototype.

WETLAND FUNCTIONS

Wetlands play numerous roles in the landscape, and many authors have listed and characterized these roles (National Research Council, 1992). By far the most fundamental is their hydrologic role. The interaction of wetlands with surface, ground, or vados waters establishes the conditions for all of the other functions, whether they are related to wildlife, water quality management, flood control, or the production of food and fiber. Further, wetlands help establish the necessary hydrologic conditions for their own survival as well as for surrounding habitats. They act to store water during high flow periods, releasing it during droughts for their own use as well as for wetlands downstream. They prevent and reduce turbidity, facilitating the growth of submerged vegetation. These and many other hydrologic functions are affected by wetlands.

Hydrology is the study of the movement of water from the atmosphere, across the land's surface, and back to the atmosphere. Throughout this cycle, numerous processes control the quantity of water that is either in motion or in storage and the pathways that the water follows (Fig. 2-2). The processes include precipitation (rainfall and snow), interception and surface storage, infiltration, percolation, soil and groundwater storage, interflow, base flow, streamflow, evaporation, and transpiration. These processes apply to all elements of a watershed: forests and prairies, grassed and paved surfaces, and streams, lakes, and wetlands. They apply in every climatic region, desert or tropical landscape. The only differences among these varied landscapes are the quantities and principal paths of storage and movement over time.

Water quality must to be appended to the definition, for water quality and quantity are inextricably tied. The rate of flow affects the degree of chemical treatment and, ultimately, the quality of water moving from one reach or body of water to another. Wetlands, in particular, not only affect the rate of flow but the quantity of water, which affects the chemical quality. Wetlands generally slow the movement of water, allowing greater opportunity for evapotranspiration and groundwater recharge, thus reducing the amount of water and increasing the opportunity for bio- and geochemical reactions to take place. These

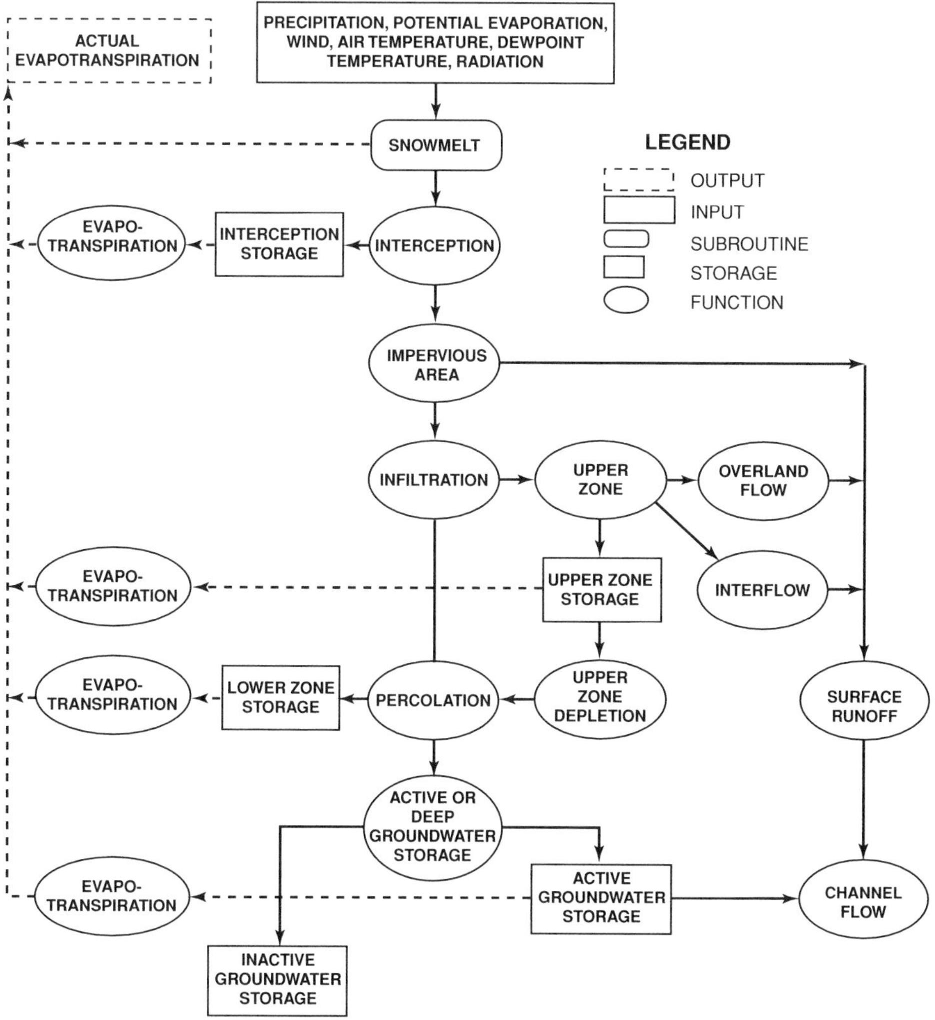

Figure 2-2 Hydrologic processes. Adapted from Hydrocomp, 1972.

reactions are further facilitated by the presence of more interactive surfaces (plant leaves, stems, and detritus), shallow depths, which promote aeration and settling, and the underlying anoxic zone, which promotes denitrification and immobilization of heavy metals, among other reactions. These hydrologic and chemical interactions must be assessed together. As observed by Kusler and Kentula (1990), "Careful attention to wetland hydrology is needed in design . . . wetland hydrology is the key (although not necessarily sufficient in itself) to long-term functioning systems."

Wetlands are not efficient hydraulic structures. On the contrary, they are reasonably good at removing water from the surface flow path. Unlike a prismatic channel, such as might be designed by an engineer, the wetted surface of a wetland is at least an order of magnitude greater. This causes increased boundary friction, slowing the movement of water. Moreover, the surface area to depth ratio of a wetland is at least two orders of magnitude greater than that of a prismatic channel; consequently, a greater percentage of the water is exposed, and ultimately lost, to evaporation and infiltration. The end result is that wetlands reduce watershed yield. In a recent study of nine watersheds in Wisconsin (Hey and Wickencamp, 1997), their yield ranged from 13.5 to 10.2 inches per year corresponding to an increase in wetlands from 2% to 20% of the land surface (Fig. 2-3). This constitutes a 24% reduction in yield relative to an 18% increase in wetlands.

Wetlands affect other hydrologic characteristics as well. For example, low flows in the Wisconsin study were shown to increase with an increase in wetlands. In order to sustain the reduction in yield, in the face of an increase in base flow, high flows were decreased with an increase in wetlands. As the range of flow (the difference between high and low flow) is reduced by the presence of wetlands, so is the frequency of flow and stage fluctuations. Using the number of excursions above the mean daily flow value equaled or exceeded 50% of the time as the representative statistic, the excursion frequency ranged from 19 to 7 per year—the greater the percentage of wetlands, the fewer the excursions (Fig. 2-4).

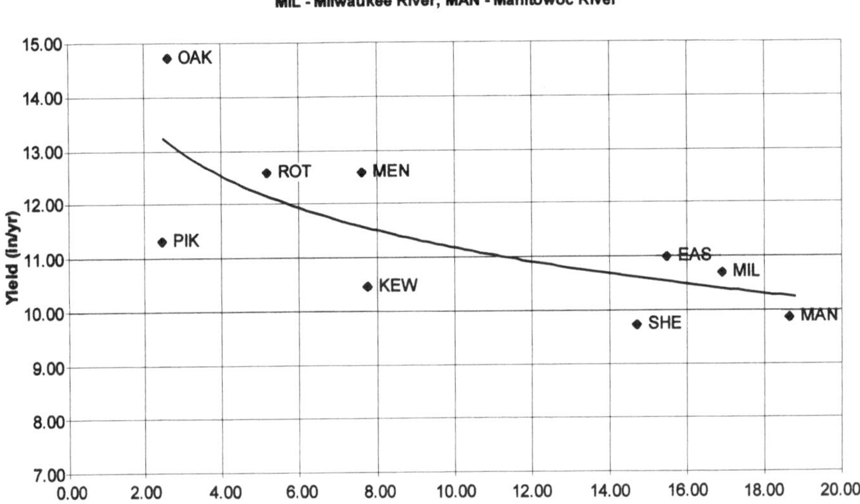

Figure 2-3 Watershed yield. From Hey and Wickencamp, 1998.

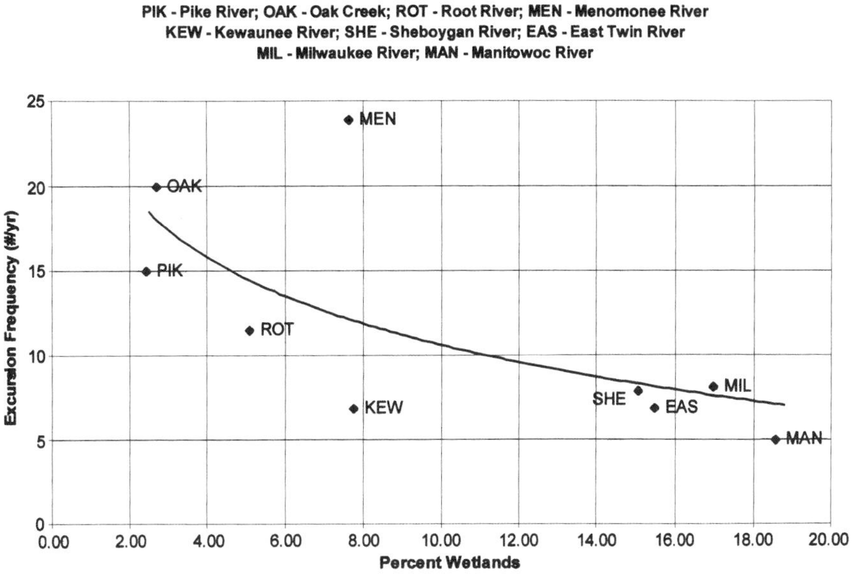

Figure 2-4 Excursion frequency. From Hey and Wickencamp, 1998.

The modulation of streamflow by wetlands can produce both positive and negative values. Although the reduction of yield may be viewed as bad, particularly from the surface-water supply perspective, the result is not all bad. Part of the loss is to groundwater, which can be accessed through wells. Higher groundwater levels and larger groundwater supplies are viewed, in many parts of the United States, as a benefit. Higher base flows provide better in-stream habitat and dilution of undesirable constituents. They also accommodate other human needs such as water supply and navigation. In fact, many smaller watersheds, which were perennial prior to agricultural and urban development, are now ephemeral because of the loss of wetlands. Before their modulating wetlands were drained, these streams supported water mills and other uses requiring year-round flow.

Native Americans understood the role of wetlands in sustaining streamflow and they recognized the role that beavers played in creating impoundments and wetlands (Morgan, 1991). In the arid west, they were reluctant to kill the animal because, Morgan concludes, these early peoples well understood the environmental consequences—loss of critical water resources and wildlife habitat. Beaver dams and the impounded water attracted a wide range of fauna and flora that were a convenient source of food for the native populations. So important was the beaver that it was embodied in their religious beliefs, which underscored the importance of the beaver and reinforced the prohibition against hunting the animal. These beliefs, in turn, preserved and sustained wetlands.

European Americans, on the other hand, decimated the beaver population, cleared and drained their control structures, and then built large, isolated reservoirs to protect against flooding and to store the increased high flows for release during low flow periods. These reservoirs, however, because of their relative large size and limited distribution, do not emulate the natural prototype.

The high level of flood damage in the United States suggests the need for the natural flood storage provided by wetlands. Despite the expenditures on traditional engineered flood control measures (i.e., levees and reservoirs), the average annual damage has increased steadily since the passage of the first Flood Control Act in 1935 (Hey and Philippi, 1995). Flood storage is a simple wetland function requiring little technical skill in its restoration. All that is required is that the storage capacity be properly positioned. Where practical, a natural wetland flood control strategy needs no emergency spillways or engineered structures. As more storage is required, the water simply spreads further upstream and across the floodplain, forming a shallow sheet of water with a minimum of destructive energy.

Given such flood storage, fluctuations in stream stage would be reduced. This has numerous advantages. Rapid stage and discharge fluctuations increase channel and bank erosion by directly entraining bed and bank materials as well as increasing the hydrostatic pressure on banks, causing them to fail ultimately. Less frequent fluctuations create more hospitable environments for the propagation of aquatic plants and animals. Substrate and plant materials are more stable, benefiting a wide variety of other aquatic organisms. These hydrologic characteristics, besides directly supporting human and wildlife needs, also support the ability of wetlands to foster chemical reactions, which affect the quality of water moving through them. The relatively long residency time and the myriad reaction surfaces offer ample opportunities for chemical and microbial transformations of the transient suspended and dissolved solids.

These interactions are affected by several factors. For example, pulses of water can shorten the detention time and adversely affect treatment efficiency. Low temperatures (near freezing) as experienced at high altitudes or in northern climates retard chemical and biological reactions. Bioturbation, such as caused by foraging carp, can resuspend solids and nutrients and increase turbidity.

In the natural landscape or in one totally contrived by humans, wetlands can have a decided but varied effect on suspended solids (Fig. 2-5). These effects depend on the incoming concentrations, the depth of the wetland, the mean residency time, and wave and other disturbances (Kadlec and Knight, 1996). Whereas suspended solids are considered conservative (they are assumed not to change their chemical form as they move through or are trapped by wetlands), other constituents such as nitrate and nitrite are altered. In the anaerobic zones of wetlands, bacteria strip the oxygen from the nitrate and nitrite molecules, releasing nitrogen gas to the atmosphere. The reductions in nitrogen can be as great as those for suspended solids, as shown in Figure 2-6. The biochemical reactions, however, change with temperature. During warmer months, more NO_3 is converted to nitrogen gas and the oxygen is consumed, as shown

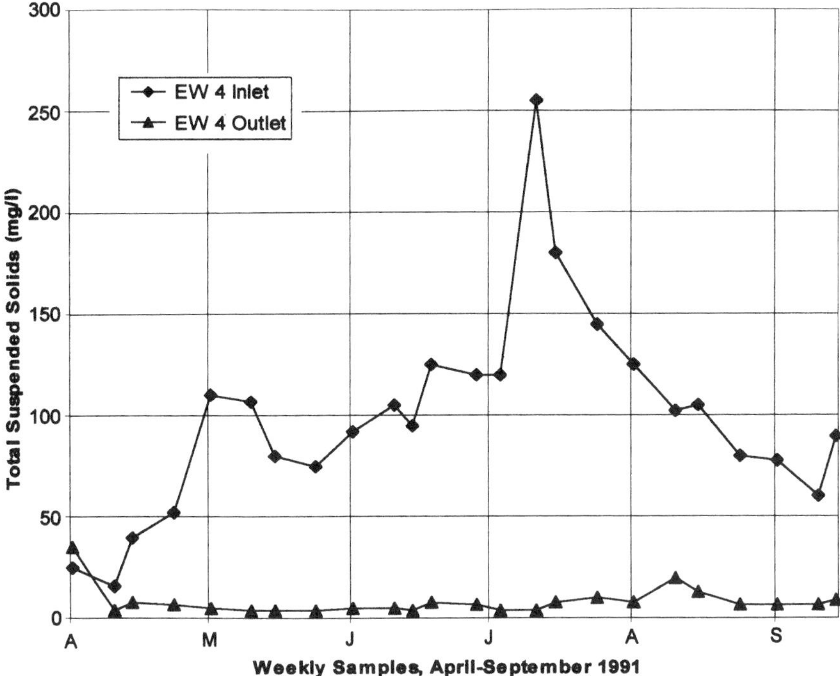

Figure 2-5 Suspended solids reduction by experimental wetland 4 at the Des Plaines River Wetlands Demonstration Project. Reproduced by permission of Wetlands Research, Inc., Chicago, Illinois.

in Figure 2-6. During the winter, in regions of the United States where temperatures fall below freezing, the microbial reactions slow and the reduction of NO_3 is curtailed. Similar chemical and physical reactions affect the other nutrients and a wide range of other organic contaminants, such as the herbicide atrazine (Fig. 2-7).

To one degree or another, the carbon cycle is affected by wetlands (Mitsch and Wu, 1995). Wetland plants, like other plants, remove CO_2 from the atmosphere and temporarily store it as living biomass. As the plant dies and decomposes, however, the carbon can be returned to the atmosphere or it can be sequestered on a more permanent basis. In shallow marshes, such as sedge meadows that dry out during periods of the year, a majority of the carbon mass may be oxidized and returned to the atmosphere. Wetlands that have sustained surface water tend to retain carbon. Bogs and deep marshes are examples in which carbon is stored, in the form of peat.

Underneath the blanket of water, or in saturated conditions, biogeochemical actions begin the process of forming hydric soils. Landscape position can be far more important in deciding the character of soils than any other factor. In

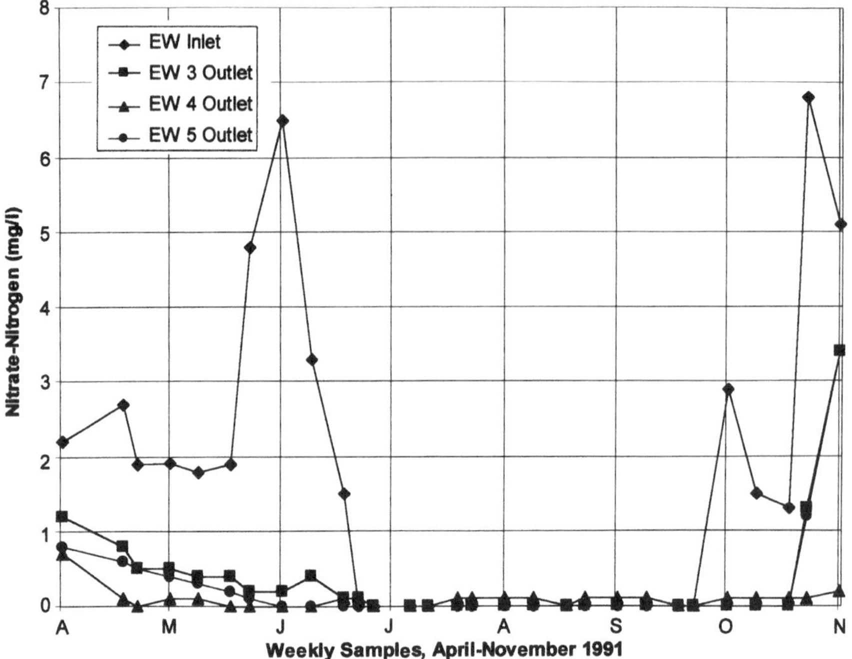

Figure 2-6 Nitrogren reduction by three experimental wetlands at the Des Plaines River Wetlands Demonstration Project. Reproduced by permission of Wetlands Research, Inc., Chicago, Illinois.

mountainous regions, soils tend to be coarser grained, containing cobbles and gravel, owing to the greater energy and carrying capacity of the stream; in flatter terrain the particle sizes are much smaller. Wetlands along coastal waters usually contain the finest materials of all, having been ground to very small particles during their journey to the sea. The character of the underlying soils depends on the parent material, whether sandstone, limestone, or granite, and the period and duration of inundation. Regardless of the responsible factors, hydric soils can form very quickly. The characteristic processes, such as chemical reduction (e.g., ferric oxide is converted to elemental iron appearing as rust-like flecks in the soil profile) begin to work. Within 5 years, the indicators of hydric soils can be observed (Vepraskas et al., 1995).

Over the soil, adjacent to and in the water, hydrophytes grow and prosper. In association with each other and their physical surroundings, they form the habitat structure of the natural landscape. Wetland plants vary by landscape position, climate, geology, and hydrologic conditions. Water depth and duration of inundation are perhaps the most important determinants of plant community. Within climatic zones, regional lists of hydrophytes and their habitat requirements are widely available.

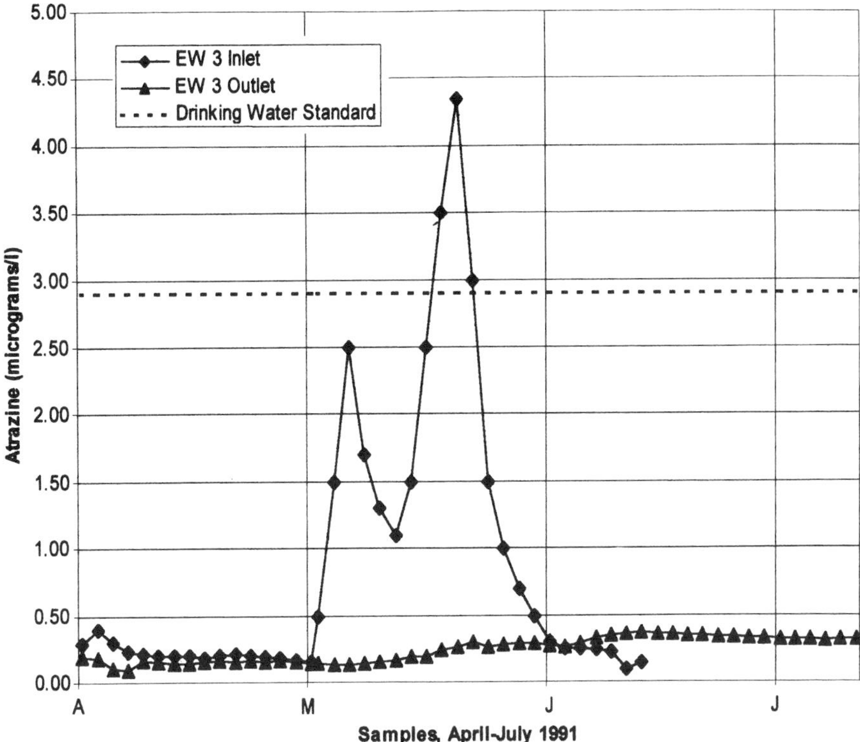

Figure 2-7 Atrazine reduction by experimental wetland 3 at the Des Plaines River Wetlands Demonstration Project. Reproduced by permission of Wetlands Research, Inc., Chicago, Illinois.

CONCLUSIONS

The hydrologic, chemical, and biological reactions of wetlands vary only in degree from one landscape position to another. On the side of a mountain slope, a fen and the associated microbial populations act to alter the flow regimes and chemistry of the sustaining stream of water. In intertidal and tidal basins, the ebb and flow of fresh water and saltwater have diurnal effects on the depths and surface areas, but nonetheless support the growth of microbial communities that interact with the surface and interstitial waters. The creation or restoration of a wetland must combine these controlling factors in accordance with the type of wetland desired, the landscape setting, and the ambient climate—a task that may be easier said than done.

Most restoration projects in the United States are small-scale—from 1 to 2 acres to perhaps 15 to 20 acres. The controlling factors can be applied or manipulated without fear of adversely affecting nearby vested uses, except perhaps for backing water up on someone else's property. With larger-scale projects, applied to a watershed for example, and with projects intended to provide specific benefits, two concerns must be addressed:

1. How much wetland area is required to satisfy a given set of objectives?
2. Where should the wetlands be placed within the watershed and relative to each other?

Without specific details, only a general response to the first question can be offered. The greatest effects of restoring wetlands in the Midwest seem to occur when wetlands represent between 0% and 10% of the watershed area (Johnston et al., 1990; Hey and Wickencamp, 1998). Beyond this range improvements still occur but at a much lower rate. Beyond 10%, the law of diminishing returns takes effect—yield, peak flows, and excursion frequencies are reduced to a lesser extent by each additional acre of wetland. Similarly, base flow is increased by a lesser extent with each additional wetland acre.

The answer to the second question is quite simple: as close to the point of need as possible. Ogawa and Male (1983), studying flood control on the Charles River in Massachusetts, clearly demonstrated and articulated the proximity rule. The further away the controlling storage is, the greater the opportunity for intervening flood flows to negate the hydraulic benefits of the lower stages brought about by the storage structure. Similarly, reducing biochemical oxygen demand far upstream from the point requiring sustained high levels of oxygen only permits the intervening demands to consume the extra oxygen provided by the upstream treatment capacity.

In addition to the issues of proximity and scale, each restoration project should be viewed as potentially contributing to the solution of downstream, larger-scale problems. Wetlands were removed from the natural landscape acre by acre causing the gradual, incremental degradation of the aquatic environment. Their return, in the same fashion, acre by acre, will result in the gradual, incremental improvement in the environment.

Flood damage could be greatly reduced if wetlands occupied the floodplain rather than farms, towns, and industrial facilities. In 1993, the floodwaters that ravaged the upper Mississippi Basin could have been harmlessly and productively stored on 13 million acres of wetlands. This area, in addition to the existing wetlands, would bring the total to 32 million acres in the entire watershed, which would represent 7% of the watershed area and only 60% of the presettlement wetlands (Hey and Philippi, 1995).

Wetlands of all varieties along the waterways and in upland areas of the Mississippi watershed could help address the nitrate and silicate imbalance in the river's discharge to the Gulf of Mexico. The imbalance of these chemicals is thought to be causing the depletion of dissolved oxygen over a 7,000-square-mile area, affecting shellfish and other aquatic life (Rabalis, 1996).

Project by project, wetlands can be reestablished on the landscape, natural or otherwise. As wetlands flourish and landscapes emulate the natural prototype, wildlife will proliferate and many of our nation's environmental problems will be solved.

REFERENCES

Butler, D. R., *Zoogeomorphology: Animals As Geomorphic Agents*, Cambridge University Press, New York, 1995.

Chumbley, C. A., R. G. Baker, and E. A. Bettis III, *Midwestern Holocene Paleoenvironments Revealed by Floodplain Deposits in Northeastern Iowa*, American Association for the Advancement of Science, Iowa City, IA, 1990.

Cowardin, L. M., V. Carter, F. C. Golet, and E. T. LaRoe, *Classification of Wetlands and Deepwater Habitats of the United States*, FWS/OBS-79/31, Washington, DC, 1979.

Dahl, T. E., *Wetlands Losses in the United States: 1780s to 1980s*, U.S. Department of the Interior Fish and Wildlife Service, Washington, DC, 1990.

Davis, C. A., *Peat: Essays on Its Origin, Uses, and Distribution in Michigan*, Wynkoop, Hallenbeck, Crawford, Co., Lansing, MI, 1907.

Desor, E., *La Foret Vierge et le Sahara*, Librarie Sandoz et Fischbacher, Paris, 1879.

Foreman, G., *Adventure on the Red River*, University of Oklahoma Press, 1937.

French, B., Personal Correspondence, Bat Conservation International, Austin, TX, 1998.

Frost, S. L. and W. J. Mitsch, *Resource Development and Conservation History Along the Ohio River*, School of Natural Resources, Ohio State University, Columbus, OH, 1989.

Hajac, E. R. and G. R. Walz, *Evidence of Prehistoric Beaver Activity in the Des Plaines River Valley*, Wetlands Research, Inc., Chicago, IL, 1998.

Hey, D. L. and N. S. Philippi, "Flood Reduction Through Wetland Restoration: The Upper Mississippi River Basin As a Case History," *Restoration Ecology*, Volume 3, Number 1, 1995.

Hey, D. L. and J. A. Wickencamp, "Effects of Wetlands on Modulating Hydrologic Regimes in Nine Wisconsin Watersheds," *Water Resources and the Urban Environment*, E. D. Loucks, editor, The American Society of Civil Engineers, Reston, VA, 1998.

Hydrocomp, Inc., *Hydrocomp Simulation Program*, Palo Alto, CA, 1972.

Johnston, C. A., N. E. Detenbeck, and G. U. Neimi, "The Cumulative Effect of Wetlands on Stream Water Quality and Quantity,"*Biochemistry*, Volume 10, 1990.

Kadlec, R. H. and R. L. Knight, *Treatment Wetlands*, Lewis Publishers, Boca Raton, FL, 1996.

Kusler, J. A. and M. E. Kentula, *Wetland Creation and Restoration: The Status of the Science*, Island Press, Washington, DC, 1990.

McCulloch, D. and D. Hopkins, *Evidence for an Early Recent Warm Interval in Northwestern Alaska*, Geological Society of America, Menlo Park, CA, 1966.

Mills, E. A., *The Beaver World*, Houghton Mifflin Co., Boston and New York, 1913.

Mitsch, W. J. and X. Wu, "Wetlands and Global Change," *Soil Management and Greenhouse Effect*, Lewis Publishers, Boca Raton, FL, 1995.

Morgan, G. R., *Beaver Ecology/Beaver Mythology*, University of Alberta, Edmonton, Alberta, 1991.

Naiman, R. J. and C. A. Johnston, "Alteration of North American Streams by Beaver," *Bioscience*, December, 1998.

National Research Council, *Restoration of Aquatic Ecosystems*, National Academy Press, Washington, DC, 1992.

Ogawa, H. and J. W. Male, *The Flood Mitigation Potential of Inland Wetlands*, Water Resource Research Center, University of Massachusetts, 1983.

Rabalis, N. N., *Mississippi River Water Quality: Status, Trends, and Implications*, Environmental State of the State Conference, Baton Rouge, LA, 1996.

Ruedemann, R. and W. J. Schoonmaker, "Beaver-dams: Dams As Geological Agents," *Science*, Volume 88, Number 2292, 1938.

Van der Valk, A., *Northern Prairie Wetlands*, Iowa State University Press, Ames, IA, 1989.

Vepraskas, M. J., S. J. Teets, J. L. Richardson, and J. P. Tandarich, *Development of Redoximorphic Features in Constructed Wetland Soils*, Wetlands Research, Inc., Chicago, IL, 1995.

CHAPTER 3

THE DRAINING OF THE WETLANDS

As European settlers moved west across the North American continent, their livelihood depended on their ability to grow crops. Early pioneers were homesteaders, carving small plots out of the wilderness to grow food for themselves and their animals. The establishment of self-sustaining farming operations was necessary to their survival. Soil fertility was only one criteria for success. They also needed access to forests for building materials and fuel and a source of water. The forests of the East provided these necessities but as they moved into Ohio and the Great Lakes region, they were confronted with vast wet prairies that were, in the earliest period of pioneering, the most undesirable locations for settling. They preferred locations near or in the forests that provided them with the wood, the water, and the protection from hostile Indians that the exposed prairies denied them. Moreover, they were poorly equipped to work the heavy wet soils of the vast Midwestern prairies.

> The open, rolling prairie lands offered little attraction to early settlers. Although the low, luxuriant growth attested to good soil, for many months of the year these lands were wet and soft. The black, sticky soil could not be turned by the iron and wooden ploughs that had proven effective in the sandy loams from which the emigrants had come. Gnats, flies, snakes, and wild animals infested the tall grasses, and the dread black swamp fever was thought to steal out of these places at night to take toll of settlers and their families.
>
> —Conger, *History of the Illinois River Valley*, 1932 (through Hannah, 1960)

Wet soils had been an impediment to farming where they were found on the East Coast, but the swamplands from Ohio westward into the Mississippi valley presented a challenge both quantitatively and qualitatively in excess of any that early travelers had experienced before. Firsthand accounts described the conditions:

> The distance from this [the Ohio state line] to Fort Wayne is 24 miles, without a settlement; the country is so wet that we scarcely saw an acre of land upon which settlement could be made. We traveled for a couple of miles with our horses wading through water, sometimes to the girth. Having found a small patch of esculent grass (which from its color is known here as bluegrass) we attempted to stop and pasture our horses, but this we found impossible on account of the immense swarms of mosquitoes and horse flies, which tormented both horses and riders in a manner that excluded all possibility of rest.
>
> —W. H. Keating, University of Pennsylvania, traveling in 1823

> During four consecutive weeks there was not a dry garment in the party, day or night. Consider a situation like the above, connected with the dreadful swamps through which we waded, and the great extent of windfalls over which we clumb and clambered; the deep and rapid creeks and Rivers that we crossed, all at the highest stage of water; that we were constantly surrounded and as constantly excoriated by swarms or rather clouds of mosquitoes, and still more troublesome insects; and consider further that we were all the while confined to a line, and consequently had no choice of ground . . . and you can form some idea of our suffering conditions.
>
> —Harry A. Wiltse, 1847, on a survey expedition in Wisconsin (Prince, 1997)

Even Charles Dickens found the prairies aesthetically unappealing:

> Looking toward the setting sun, there lay, stretched out before my view, a vast expanse of level ground; unbroken, save by one thin line of trees, which scarcely amounted to a scratch upon the great blank; until it met the glowing sky, wherein it seemed to dip; mingling with its rich colors, and mellowing in its distant blue. There it lay, a tranquil sea or lake without water, if such a simile be admissible, with the day going down upon it: a few birds wheeling here and there: and solitude and silence reigning paramount around. But the grass was not yet high; there were bare black patches on the ground, and the few wild flowers that the eye could see, were poor and scanty. Great as the picture was, its very flatness and extent, which left nothing to the imagination, tamed it down and cramped its interest.
>
> —Charles Dickens, *American Notes* (Prince, 1997)

Eventually the pressures of population expansion and the scarcity of the desirable farming tracts in and near the woodlands forced settlers onto the prairies. By 1879, "the highest proportion of land growing corn and the highest yields gathered were from former wet prairies" (Prince, 1997). Once they had mastered techniques for farming this new landscape, the corn belt and its phenomenal productivity were born. It was drainage and, in particular, tile drainage that made this possible.

AGRICULTURAL DRAINAGE IN THE NEW WORLD

Drainage is at least as old as recorded history. Surface drains are known to have been used as early as 400 B.C. by the Greeks in Egypt: A plan of a ditching system dating from 250 B.C. survives on papyrus. Marcus Porcius Cato gave specific directions for draining land in the second century B.C. and Pliny and Palladius spoke of drains and drainage several centuries earlier. The Romans used both open and closed drains to remove water from the soil, often filling them with stones or bushes.

Eighteenth-century agricultural drainage activities in the United States were limited to excavation of open ditches and clearing streams. The Dismal Swamp in Virginia and North Carolina, which was first surveyed in 1763 by George Washington, had been channelized for land reclamation and water transportation by the Dismal Swamp Canal Company by the mid-1790s. The Cawcaw Swamp was drained by the colony of South Carolina in 1754. A drainage outlet for the city of New Orleans was constructed around 1794. Individual farmers along the length of the Atlantic seaboard removed excess water from their fields with open ditches and cleared brush and debris from their waterways to speed the flow of water from their land.

Subsurface drainage was less familiar to American farmers, although it had been practiced in Europe for many years. Tile drainage, which took its name from the medieval clay roofing tiles used in the earliest experiments in France, was the form that subsurface drainage took in the new world. Slow to be accepted, this new farming technology quickly transformed the wet swampy landscape of the Midwest.

A Scotsman, John Johnston, is credited for introducing tile drainage to the North American continent in the early 1800s. Drainage practices were well known in Great Britain, where they were used to reclaim up to 700,000 acres of fens on the east coast of England. Joseph Elkington, in 1764, "happening to drive an auger through the bed of a trench, discovered the existence of a water-bearing stratum, beneath, by drawing the water from which, the surface and subsoil became thoroughly drained. From this accident came into being what is known as Elkington's System of Drainage" (Weaver, 1964). Elkington's theories provided the basis for the British Parliament's commissioning the publication of "An Account of the Mode of Draining Land, Etc." by John Johnstone in Edinburgh, Scotland, in 1797. It was later published in Petersburg, Virginia in 1838.

John Johnston came to America from the Scottish borderlands in 1821 and settled in Seneca County, New York, where he bought a farm and began to raise grain—first barley, then wheat—under difficult early conditions. In 1835, he imported samples of horseshoe-shaped drainage tiles from Scotland, the first recorded in this country. Johnston credited his considerable success as a farmer to "D, C, & D"—dung, credit, and drainage—and to the grandfather that raised him, whom he quoted as saying "Varily, all the airth needs draining" (Weaver, 1964).

The tiles were first made by hand from clay rolled into sheets ½-inch thick, cut into rectangles, bent over a pole in the shape of a horseshoe, allowed to air dry, and baked in a kiln. When a workman's leg was substituted for the pole, they were called "shinbone tile." By 1838, Johnston had convinced Benjamin F. Whartenby to manufacture his tiles with a molding machine at his pottery factory in Waterloo, New York. By 1850, the Whartenby tile factory employed eight people and was producing 840,000 tiles and pipes (of various lengths) in a season. Competition quickly sprang up. In 1848, the first Scraggs machine, which made tiles by the extrusion method, was imported from England. Subsequent machines appeared, including one invented in 1851 by John Dixon, who later developed the Downdraft Inside Flue tile kiln. Although Johnston had first suffered criticism and ridicule from his neighbors, the benefits of tiles in the grain-producing county were eventually accepted, thanks to testimonials from early experimenters such as "the extent of the failure of the wheat crop the present season has, more conclusively than it has been for many seasons before, shown the utility of Tile Draining—for in no instance, and it is safe to say on no premises where lands were properly under-drained, has wheat been known to winterkill—which everyone knows is caused by the freezing and thawing of water standing upon the wheat" (Weaver, 1964).

The critical arguments, for farmers, were the economic ones. Johnston claimed that the increased productivity he achieved by laying tiles paid for the cost of drainage in 2 years. His critics countered that it cost $25 per acre, which a contemporary claimed was "rather startling to farmers in a country where, in many places, we can purchase farms with fences, buildings, and other improvements upon them, for that money" (Weaver, 1964). Yet as production methods improved, the costs went down and tile drainage gained popularity. In 1838, Whartenby provided Johnston with tile at $18 per thousand; by 1856 it was $10–12 per thousand. Johnston had installed 9 miles of tile by the end of 1850 and 72 miles by the time of his death in 1880.

The main benefits that Johnston claimed for his tile drainage operations were the following:

- Consistency—although his largest single crop was in 1833, the average yield, since then, was much higher. Tiling reduced risk.
- When the midge destroyed most of the wheat, his yield was four times that of his neighbors because of the drainage. Midges hit the fields not ripe by the twentieth of July, whereas wheat started early on well-drained land grew and ripened fast, because the cooling process of evaporation had been reduced or eliminated.
- Ease of working the land.
- Elimination of quack grass.

Johnston's role in introducing tile drainage to the United States was enhanced by his stature as a leader in the farming community. He read and wrote about tile drainage, promoted and argued its benefits, and proved himself an exceptional man as well as farmer. He was awarded prizes by the Seneca County Agricultural Society, an organization that he at one time served as president. Along with Whartenby, Johnston was joined by two other Seneca County men in spearheading the cause of subsurface tile drainage, Robert J. Swan and John Delafield. Swan employed it on a large scale on his own farm, and Delafield promoted it ceaselessly as an agricultural leader and president of the County Agricultural Society.

By 1857, the use of tile drainage was widespread locally, and one farmer, W. B. Pratt, wrote "of some thirty acres of spring grain on the writer's farm —one half would have been nearly or quite ruined but for some 1,100 rods of drainage in the same; and the present appearance is that the enhanced product this year will be nearly equal to the total cost of the drains, albeit said cost has been fully and simply compensated by former crops" (Weaver, 1964).

Once the initial resistance was overcome and tiling became accepted, tile factories proliferated. Tiles were designed to be triangular, round, oval, heart shaped, egg shaped, or square in shape; they had soles and rims and ridges and "club feet." Some were flared at one end, contracted at the other. Yet they were all costly—expensive and heavy to transport—and to keep costs down, small operations had to be located close to their use. By 1882, there were 1,140 tile factories in the United States, more than 90% of which were in the states of Indiana, Illinois, and Ohio. "Tile factories had become one of the most distinctive features of wet clay lands east of Chicago" (Prince, 1997).

In the next 20 years, hundreds of thousands of miles of tile were laid in the Midwest. What farmers were experiencing, apparently, was many or all of the 12 benefits of drainage listed in John Klippart's *The Principles and Practice of Land Drainage*, third edition, 1888 (Prince, 1997):

- Removes stagnant water from the surface.
- Removes surplus water from the undersurface.
- Lengthens the seasons.
- Deepens the soil.
- Warms the undersoil.
- Equalizes the temperature of the soil during the season of growth.
- Carries down soluble substances to the roots of plants.
- Prevents freezing out or heaving out.
- Prevents injury from drought.
- Improves the quality and quantity of crops.
- Increases the effect of manures.
- Prevents rust in wheat and rot in potatoes.

The nineteenth century was a period of population explosion, rapid expansion into the western territories, and land speculation. During the mid-1880s, the U.S. population increased dramatically (300% from 1820 to 1860). These people needed land. By the time the Homestead Act was passed, in 1862, granting a homestead not exceeding 160 acres to anyone who lived and made improvements on it for 5 years, almost all the public land in Ohio, Illinois, Indiana, and Iowa was gone. The momentum to convert wetlands to agricultural production with drainage did not stop at the corn belt, however, but moved westward along with the settlers. In California, large-scale drainage and agricultural conversions of the wet areas between the Coastal and the Sierra Nevada mountain ranges resulted in the loss, between 1870 and 1920, of the vast majority of the wetlands of the Central Valley (United States Geological Survey, 1996). Most of the Bureau of Reclamation's Irrigation Projects in the west experienced problems requiring drainage systems to be installed, including such arid locations as the Salt River Valley in Arizona, the Newlands project in Nevada, the Modesto Irrigation District in California, and Mesilla Valley in New Mexico (Beauchamp, 1987).

Most of these systems were based on subsurface tile drains with surface receiving ditches, but some were exclusively surface drains. Many, particularly those on floodplains behind levees, lake and coastal plains, peat land, and irrigated lands, were drained by pumping. They all permitted the conversion of land to agricultural production and promoted the economic expansion of the new country. This expansion was grounded not only in the increased productivity of these reclaimed lands but also in the farmers' ability to take advantage of that productivity by two important innovations: advanced drainage technology and the drainage district as a legal entity.

THE TECHNOLOGY OF DRAINAGE

Techniques for getting the water off the land became more and more efficient and less and less costly with experiments that were already underway in the middle of the nineteenth century. Landowners began with open ditches dug by hand, experimented with "mole" plows that dug smooth-sided subsurface drains, found that tile was more effective for subsurface drains, and eventually mechanized the digging of surface drains and the laying of subsurface tiles.

John Johnston's tiles had been laid by hand—women and boys were often available for the work and the costs were low. Two persons with a team of horses could trench 20 to 30 rods (1 rod equals 16.5 feet) per day. The ditches that were dug across the vast Kankakee marsh in 1870 were dug by hand. Mechanized ditch digging was faster and easier, and shovels were eventually replaced by wheel-type trenchers. A revolving wheel ditch digger was demonstrated at the New York State Fair as early as 1854. By 1860, ditching machines were being used in Illinois. Steam-powered dredges were used after 1884. And by 1867, Ohio was producing more than 2,000 miles of drain tiles per year from 500 steam-powered plants (Beauchamp, 1987).

Mole ditchers, first used in New York State in 1867, permitted smooth hard subsurface drains to be channeled out below the surface at the rate of half a mile a day with two yoke of cattle. But it became clear that tiles were more durable, didn't collapse, and didn't require cleaning, and were eventually dragged along behind to be inserted in the tunnels made by the ditcher. The tiles were laid end to end in carefully prepared trenches at about 4 feet deep; the lines were laid 40 feet apart in clay and 60 to 100 feet apart in loam. Tiles had been made from concrete as early as 1862, yet until about 1900, concrete drain tile was used only where good clay was not available. Bituminized fiber pipes began to appear in the 1940s, followed by corrugated plastic tubing. By then the technology included the construction of envelopes surrounding the drains to avoid their clogging. By the middle of the twentieth century, the Agricultural Conservation Program (ACP) required compatibility with the American Society for Testing Materials (ASTM) standards and specifications for drainage tiles before they would authorize cost-sharing payments for installing tile drainage.

Today the industry uses corrugated plastic drain tube products and synthetic drain envelope materials with mechanized handling that speeds and cheapens the process. There are two types of trenching machines: the wheel type and the ladder or chain type which, by the late 1960s, were able to lay 50 feet of drain per minute (Fouss and Reeve, 1987). Sophisticated mole plows that lay 80 to 150 feet per minute have been used since 1969, feeding the drainage tubing into the ground through the slit opening created by passing the blade through the soil, a technology made possible by the development of laser beam automatic grade control. Although the capital investment is greater for mole plows than for trenchers, the installation charge is reduced by the increase in speed. Computer software is being used increasingly for farm water management, and weather forecasts and climatic data are factored into engineering and economic modeling packages such as DRAINMOD.

THE DRAINAGE DISTRICT

The new technologies required large capital investments that were beyond the means of all but the very large landowners. During the period between 1880 and 1920, the size of farms increased and the number of farmers working land owned by others (tenant farmers) increased from 25% to 38%. The way small farmers stayed alive was by combining their efforts and their limited capital. The creation of the drainage district as a legal entity made this possible.

Between 1847 and 1872, most of the states had passed the laws establishing drainage organizations, which are authorized in 45 states today. In most of these states (39), they are called drainage districts. In many states several optional organizations are available to landowners. The two main types of organization used today are the more popular drainage districts, legally organized as separate entities, and county drains, where the districts are administered by the public officials of the county. The laws established the way of organizing, the method of allocating costs, the authority to levy and collect taxes, the right of condemnation, and the method of financing by sale of bonds.

The drainage district is a "quasi" local government, having fewer powers than municipalities and the inability to levy general taxes. Their powers are limited to those that will allow them to construct drainage works and make a distribution of the costs equitable in proportion to the benefits to participating landowners. They are able to own interest in real and personal property, to make contracts, borrow money, issue bonds, pay debts, sue and be sued, levy and collect taxes, and adopt regulations. The drainage districts are run by a board of commissioners, usually elected but sometimes appointed. Drainage improvements are financed by assessments that, in several ways, attempt to proportion costs according to benefits received. In addition to drainage districts, some states have created conservancy districts whose functions and responsibilities extend beyond drainage and whose jurisdiction may encompass wide areas.

Organized drainage projects proliferated between 1870 and 1920, draining more than 48 million acres. In some cases they had overextended themselves financially, however, and the creation of new districts slowed with the onset of World War I. After the 1929 crash, many of the weaker drainage districts collapsed, and by 1940, the number of acres drained by organized drainage districts had declined to 40 million acres. With the end of World War II, however, drainage district activities expanded and, between 1945 and 1965, another 24 million acres was put under their control, bringing the total to 64.33 million acres. Although the jurisdiction of drainage districts continued to expand, at a slower rate, between 1965 and 1975, the percentage of total drained land within these organizations continued the decline that had begun with the onset World War I. Whereas in 1900, 99% of drained land was in drainage organizations, by 1965 that percentage had declined to 69% and by 1985 to only 59%, with the remaining 41% of drained acreage owned by independent farm systems (Pavelis, 1987). From the 1975 peak of 66 million acres, the percentage of drained agricultural land that was controlled by drainage districts had dropped from 64% to 59% by 1985, but the actual amount of land controlled by the districts had diminished only slightly, from 66 million to 65.3 million acres. During this same period there were 7 million new acres drained on the independent farms outside the districts.

It was the organizational mechanism of the drainage district that made it possible for American farmers to generate the capital necessary to take full advantage of the new drainage technologies. Thus was created the highly productive corn belt across a region that had been considered, upon first viewing by European travelers, to be untillable. By 1879, "the highest proportion of land growing corn and the highest yields gathered were from former wet prairies," and by 1920, "all but a few patches of wet prairie soils lay within the boundaries of organized drainage enterprises" (Prince, 1997).

DRAINAGE LAW

The rapid expansion of population and land holdings created the need for a body of drainage law to adjudicate differences among landowners within the same drainage basins. The first colony-wide drainage law was passed in New Jersey in 1772, and Maryland had drainage laws as early as 1790. Other states followed. One legal scholar observed that "drainage law didn't matter as long as it was cheaper to buy a new farm than to drain the farm one already owned. It was not until the rich prairie lands of [Illinois] could no longer be had for the asking . . . that the leading case of Gilman vs. Madison Co. R.R. in 1869 fixed and determined the rights of drainage in Illinois" (Hannah, 1960).

Drainage law revolved around three general principles that were not particularly consistent among themselves: the civil law rule, the common enemy rule, and the reasonable use rule. According to the first, also called the natural flow rule, a landowner lives with the advantages or disadvantages that nature (including water flows) has brought him, and he may not obstruct that natural flow to the detriment of others. The common enemy rule recognizes water as the enemy that a landowner has the right to control. Finally, the reasonable use law allocates to the landowner the right to deal with the problems that surface water creates, as long as he acts reasonably. Most authorities agree that these rules do not provide for easy solutions to drainage conflicts. One scholar declares them all to be unreasonable, pointing out that the courts rather than the legislatures have written the only usable drainage laws (Hannah, 1960). Another is quoted as saying that "all of the common-law rules involve uncertainty; under the natural flow an owner has to show that the lower owner obstructed the patterns; under the common enemy, if everybody gets rid of their waters they may come into conflict with their neighbors all having the same rights; and the reasonable use test is uncertain because it means reasonable in the larger setting" (Sandretto, 1987).

In Illinois, for example, the right to drain is strongly supported by the laws (Uchtmann and Gehris, 1997). The courts give the dominant tenement (the upstream owner) the right to change the amount of water drained and the speed of its flow, as long as it drains into its natural basin. Overflow waters from a creek are considered to be surface waters and therefore owners of lower lands are bound to receive them. Only if the flow is "unreasonably" increased for reasons unrelated to *good husbandry* is the owner of the higher ground liable for damage to the lower land. Both the lack of good husbandry and the damage must be proven by a plaintiff. The owner of lower land (the servient tenement) may not obstruct the flow of surface water by building a dam or levee or anything else if it interferes with the drainage of higher land. In fact, the civil law specifies that willful and intentional interference by an owner of lower land is considered to be a petty offense and is subject to a fine. And although the owner of higher land can't compel the owner of lower land to clear out the ditches, in some circumstances the owner of higher land has the right to enter the servient tract to make reasonable repairs.

If, on the other hand, owners of higher ground fail to take action when owners of lower land dam or obstruct the flow of surface water, the servient owners have thus acquired a right to maintain the dam or obstruction by what is known as "prescription" or "prescriptive use." The period of such use in Illinois is 20 years, and by the same process owners of higher land may acquire the right to change the place where their surface water enters lower ground. Whether an owner has acquired such a right is a difficult question to resolve.

THE ROLE OF THE GOVERNMENT

As the development and expansion of subsurface drainage in the United States grew, so did the role and responsibilities of government in relation to it. State and local governments sponsored drainage projects of their own as early as the eighteenth century, but it was not until the middle of the nineteenth century that the federal government became a major player, indirectly, in agricultural drainage activities, upon passage of the Swamp Land Acts.

Early interest in drainage was primarily for land reclamation and secondarily for public health. Local and colonial (later, state) governments in Delaware, Maryland, New Jersey, Massachusetts, South Carolina, and Georgia supported and participated in a variety of drainage activities similar to the early Cawcaw and Dismal Swamp projects during the eighteenth and nineteenth centuries. New York City's Central Park was drained in 1858, primarily for public health benefits. Malaria was prevalent in the early states and, although the mosquito had not yet been identified as the carrier, the disease was associated with wet areas and swamps.

The Swamp Land Acts of 1849, 1850, and 1860 were the first important federal actions related to agricultural drainage. The federal government owned vast amounts of swampland along the Mississippi and other Midwestern rivers. Their disposition was controversial and, after more than 20 years of debate in Congress, the first of the acts was passed that eventually transferred millions of acres of "swamp and overflow land" to 15 states, on the condition that they would be resold and the funds from their resale would be used to drain and thus reclaim them for cultivation. The congressional debate over transfer of these lands to the states was won by the argument that the states were indeed willing and able to drain these lands (Prince, 1997). The first step was for the federal government to locate and describe legally these lands, a difficult surveying task at best (see Wiltse's quote above).

The first federal land survey was established by the Land Ordinance of 1785, drawn up by a congressional committee chaired by Thomas Jefferson. Jefferson's goals for the survey were (1) to bring the lands northwest of the Ohio River and south of the Tennessee River under the control of the federal government, (2) to sell these lands in order to reduce the federal debt, and (3) to ensure that the boundaries were clearly marked and accurately registered. The 1785 ordinance was replaced by another in 1796, and further amendments were added in 1815. The surveys made possible the sale of lands, and timberland quickly passed from the federal government to private holders. Only portions of the less popular prairies, along with the wetlands and overflow lands (flood plains), had been slow to sell.

In the execution of the Swamp Land Acts the states argued over the appropriate delineation of lands—what was swamp and overflow land and what was not—and after 1854 they made their own selection of property to be transferred. There were accusations of fraud, yet states were allowed to continue to submit claims up to the last quarter of the nineteenth century, and claims continued to be disputed into the first quarter of the twentieth century.

The process resulted, in the end, in the transfer of 64 million acres to the states, mainly to the Mississippi River Basin states, where the largest grant of more than 9 million acres went to Louisiana. In addition, Florida received the largest grant of all, 20 million acres, and California received just over 2 million acres. Florida, Illinois, Iowa, and Missouri directly assigned the lands to counties. By 1868, half of these land grants were in the hands of speculators; and "apart from Louisiana which gave high priority to constructing levees and digging drains, no other state honored its obligations to protect flood plains, reclaim land for agriculture, or drain malarial swamps" (Prince, 1997). Nor did they invest the proceeds from land sales in drainage projects, which would have been far more costly than the proceeds from the sales could cover. Minnesota, for example, used the money to build insane, blind, deaf, and dumb asylums and prisons. The public bodies, short on funds, counted on the private landowners to drain and reclaim the land in their own economic self interest.

Throughout the nineteenth century, land drainage was deemed to be in the broad public interest. In the words of a Minnesota Supreme Court decision, "the fact that large tracts of otherwise waste land may thus be reclaimed and made suitable for agricultural purposes is deemed and held to constitute a public benefit" (Prince, 1997). It was not until passage of the Reclamation Act of 1902 that the federal government became actively involved in land drainage. This act established a drainage specialist position and staff in the USDA's Bureau of Agricultural Engineering (BAE). In 1935, the Reconstruction Finance Corporation was established by Congress to refinance drainage and irrigation districts in distress and the BAE became responsible for 46 Civilian Conservation Corps (CCC) camps involved in the rehabilitation and reconstruction of drainage improvements organized under drainage law. In 1938, these camps and other drainage activities were assigned to the Soil Conservation Service (SCS). Even in 1941, drainage and irrigation work was viewed by the USDA as "conservation practices" to be included in farm conservation plans. The Flood Control Act of 1944 authorized the U.S. Army Corps of Engineers (Corps) to construct major drainage outlets and flood control channels, and in 1954, the Watershed Protection and Flood Prevention Act, PL83-566, authorized the USDA to improve drainage outlet channels and give technical assistance on drainage. Some of these functions are today assigned to the National Resource Conservation Service (NRCS), SCS's successor.

AGRICULTURAL DRAINAGE IN THE TWENTIETH CENTURY

In 1985, the total area of drained rural land was 110 million acres, of which 70% was cropland (Pavelis, 1987). Sixty-six percent of this area was drained with surface ditches, 35% with subsurface tiles. In 22 states there are 1 million acres or more of rural land that has been drained: the most in Illinois, with 9.8 million acres; then Indiana, with 8.1 million. Ninety-five percent of all drained cropland was within 23 states in 1985, averaging about 25% of their cropland drained. Today in the Midwest, there are more miles of public outlet ditches and drains than there are miles of public highways (Hooten and Jones, 1955). Of the total area drained, the actual uses of that land are estimated to have been 69% for crops, 12% pasture, and 16% woodlands, with the remaining 3% in miscellaneous uses (Pavelis, 1987).

Since 1960, the two major trends have been toward more independent farm installations than drainage districts (see above discussion) and toward more subsurface drainage than surface drainage. Subsurface drainage has been stimulated by improved equipment and material, lower maintenance costs, and minimum land loss when drains go underground.

The cost to drain land has decreased since the beginning of the century, and the farmers' annual investment in drainage has declined to a fairly low but steady level. In 1985, the cost was $225 an acre for organizational drainage and $140 an acre on individual farms, down from $345 and $225 (in 1985 dollars), respectively, in 1900. Subsurface drainage costs in 1985 were $415 per acre, only about half the cost per acre in 1965 (Pavelis, 1987). At the turn of the century drainage enterprises were investing in drainage at a rate of about $321 million per year in 1985 dollars. This investment peaked at $450 million per year between 1905 and 1910, dropped back to $300 million between 1910 and 1930, stayed low during the depression and World War II, and then recovered to $75 million per year after 1945. Between 1981 and 1985, investments in drainage fell to $50 million per year.

Pavelis (1987) has attempted to relate these costs to the benefits accrued in production increases, using a methodology that reflects only the increases in productivity to lands that would probably have been farmed anyway and excluding the land that would not have been farmed at all without drainage. For the 1982 crop year, he shows overall production increases on drained lands to be 67% higher than on undrained lands, providing him with a production index of 167 for crop sales. The equivalent index that reflects the effect of drainage on land values is 127. He acknowledges the limitations of his methodology and admits that "the economic feasibility of drainage must be determined on a case by case basis" (Pavelis, 1987). He also looked at the benefit/cost ratio of the added-on farm benefits to drainage field installation dollars in individual years and concluded that it was 1.4 in 1978 and 1.3 in 1982, but only 0.9 in 1987 and 0.75 in 1986, owing to falling land values while drainage costs were increasing. During this period, farm real estate for the counties under review had gone from $935 in 1978 to $1360 in 1982 to $920 in 1986. The argument for the economic advantages of farm drainage, though it may be difficult to prove with the limited data available, is made very convincingly by the substantial investment that farmers have made in it during the past century and a half.

THE ROLE OF NAVIGATION AND FLOOD CONTROL INTERESTS IN SURFACE DRAINAGE

Although the major modifications to the wet soils encountered by early settlers were made in the interest of land reclamation—converting the use of that land to agricultural purposes—other national interests played a role as well. It was the improvements in surface drainage within our major river systems, made primarily in the interest of navigation, that allowed those rivers to collect and carry the outflows from the subsurface drains and contributed to the overall drainage of the landscape. Eventually, however, these agricultural and navigational channel improvements increased the volume of the high flows and provided an additional incentive for river improvements—flood control.

The Constitution of the United States did not give clear authority to the federal government to participate in navigation projects, much less flood control. At the end of the eighteenth and beginning of the nineteenth centuries, the responsibility for internal waterways was considered to be a state responsibility, although Albert Gallatin, Jefferson's Secretary of the Treasury, submitted a report recommending federal aid to a great system of roads and canals to link the Atlantic Ocean to the interior of the country. The navigational activity of the Corps, between 1802 and 1823, was primarily on river and harbor projects on the Atlantic Ocean. The year 1824 was the turning point, with the Supreme Court decision in *Gibbons* vs. *Ogden* that assigned responsibility for inland waterways to the federal government, in the interest of interstate commerce. In that same year Congress passed the first Rivers and Harbors Bill to improve navigation on the Ohio and Mississippi rivers by removing sandbars, snags, and other obstacles. It also passed the important General Survey Bill, signed by President James Monroe in 1824, authorizing surveys to be made for both land routes and canals of national importance. To implement the surveys, Monroe appointed a Board of Internal Improvements that proceeded to do a series of canal surveys and, at the same time, considered river obstructions that needed clearing. The 365-mile Erie Canal, a New York State project, was finally completed in 1825, and the Board explored possibilities for other canals up and down the Atlantic seaboard and as far west as Indiana. The Federalists and the American Whig Party supported a grand system of federally supported canals, lighthouses, river and harbor works, and other navigation projects, whereas the Democrats (first known as the Democratic Republicans) maintained that such activity was inappropriate for the federal government to engage in. Gradually the nation lost interest in canals; some completed canals were actually abandoned, and in 1838, the Survey Act was repealed. The debate continued over the extent to which the federal government should participate in navigation projects, was laid aside during the Civil War, and was finally won by the pro-development interests. In 1887, according to one historian (Shallat, 1991), Congress gave the Corps "the first in a series of lavish river and harbor appropriations." This series of projects deepened and cleared rivers of obstructions, built locks and dams, constructed levees, even worked on a plan to clear mud from the Mississippi Delta.

In 1861, a *Report on the Physics and Hydraulics of the Mississippi River* by Andrew A. Humphreys and Henry L. Abbot recommended the program of levee building on the Mississippi River that was subsequently endorsed and initiated by the Mississippi River Commission, created in 1879. Although floods were already causing damage on the Mississippi River in the second half of the nineteenth century and the main projects were concentrated on the lower half of that river, the authorizing legislative language always declared the purpose to be navigation rather than flood control. In 1896, Congress authorized a study of reservoir sites in Colorado and Wyoming and, in 1892, the Reclamation Service was established in the Department of the Interior that made irrigation in the West a national policy.

Although legislation serving partial flood control purposes had been passed in 1917 and 1928, flood control was not officially acknowledged by Congress to be a nationwide responsibility of the federal government until passage of the Flood Control Act of 1936, which authorized the construction of more than 200 flood control projects across the country. Since then, the Corps has improved—cleared, channelized, and/or dredged—thousands of miles of navigable interior rivers, for flood control and navigation, all of which provide a land use conversion purpose as well in that they provide the receptacle and ultimate conveyor of agricultural drainage.

SUMMARY

In the rapid and energetic expansion of European settlers into the new lands of the North American continent, water played two major roles, aided and abetted by the U.S. government.

First, it was a constant impediment to the intensive agricultural development that was occurring, particularly in the states along the Atlantic seaboard, the Gulf coast, and the Mississippi River, in the upper Mississippi basin, and in California. In these areas wetlands were reclaimed and converted to agricultural uses through subsurface tiling and drainage systems, surface ditches, and pumping. The federal Swamp Land Acts enhanced these efforts dramatically by transferring millions of acres of overflow and swamplands to the states for the explicit purpose of drainage and reclamation.

Second, the system of interior rivers in the country was seen as essential to interstate commerce and economic expansion westward, prior to the railroads, and there were substantial federal investments, beginning in the early nineteenth century, in clearing, channelizing, and dredging the major navigable rivers of the country.

The enlarged flood flows produced by the faster and more efficient passage of water through the subsurface drains and improved surface channels produced increasing damages to the economic development on floodplains, particularly on the lower Mississippi. Congressional response to these damages was, in 1936, to endorse flood control as a federal responsibility and initiate expanded river improvement activities, all of which served to reclaim additional lands for agricultural and other economic activities.

By 1985, 110 million acres of rural land had been drained for agricultural purposes, 64 million acres of which had been granted to the states by the federal government in the Swamp Land Acts, for the specific purpose of draining. Not at all coincidentally, during the period between 1780 and 1980, wetland losses in the United States are estimated to have been approximately 117 million acres (National Research Council, 1992). The federal government, hesitant at first to play an active role in drainage activities, had eventually played the major role. As it began to appreciate the importance of wetlands, it would have to reverse that role.

REFERENCES

Beauchamp, K. H., "A History of Drainage and Drainage Methods," *Farm Drainage in the United States: History, Status, Prospects*, Economic Research Service, U.S. Department of Agriculture, Washington, DC, 1987.

Fouss, J. L. and R. C. Reeve, "Advances in Drainage Technology: 1955–85," *Farm Drainage in the United States: History, Status, Prospects*, Economic Research Service, U.S. Department of Agriculture, Washington, DC, 1987.

Hannah, H. W., "History and Scope of Illinois Drainage Law," *The University of Illinois Law Forum*, Volume 1960, Number 2, Urbana, IL, Summer, 1960.

Hooten, H. H. and L. A. Jones, "The History of Our Drainage Enterprises," *Water: the Yearbook of Agriculture, 1955*, U.S. Government Printing Office, Washington, DC, 1955.

National Research Council, *Restoration of Aquatic Ecosystems*, National Academy Press, Washington, DC, 1992.

Pavelis, G. A., "Economic Survey of Farm Drainage," *Farm Drainage in the United States: History, Status, Prospects*, Economic Research Service, U.S. Department of Agriculture, Washington, DC, 1987.

Prince, H., *Wetlands of the American Midwest*, University of Chicago Press, Chicago, 1997.

Sandretto, C., "Drainage Institutions," *Farm Drainage in the United States: History, Status, Prospects*, Economic Research Service, U.S. Department of Agriculture, Washington, DC, 1987.

Shallat, T., "Water and Bureaucracy: Origins of the Federal Responsibility for Water Resources, 1787–1838, Historical Analysis and Water Resources Development," *Natural Resources Journal*, Winter, 1991.

Uchtmann, D. L. and B. Gehris, *Illinois Drainage Law*, University of Illinois at Urbana-Champaign, 1997.

United States Geological Survey, *National Water Summary on Wetland Resources*, U.S. Geological Survey, Washington, DC, 1996.

Weaver, M. M., *History of Tile Drainage*, M. M. Weaver, Waterloo, New York, 1964.

CHAPTER 4

APPRECIATION OF WETLAND VALUES

The only good wetland, in nineteenth century America, was a drained wetland, and the drainage of wet soils and swamps was seen as a fitting and proper role of government. Economic development was in the public interest and drainage was necessary for an expanding agriculture. As they moved into the twentieth century, however, the public began to notice and become concerned about the loss of some of the natural benefits that wetlands had provided. As each new problem area was recognized as a national concern, a different federal department was given the responsibility for its solution, spreading the authority for what was ultimately seen as wetland protection among a variety of public agencies.

The first national awareness of the interrelatedness of the different features in a watershed came in the late nineteenth century with the notion that the loss of forests, at the hands of the timber barons, was resulting in increasing floods and reduced water supply. Hydrologic arguments put forward the planting of forests as an alternative to levees to prevent flooding, and forest preservation was adopted as a national goal. In the 1930s Morris Cook's *Little Waters*, which argued that disastrous floods originated in the wetland losses and erosion occurring at the far reaches of the river basin, was widely read and influenced the way people thought about flood control. At about the same time, the nation's sportsmen became concerned about the reduced numbers of migratory waterfowl; the loss of habitat and feeding areas was blamed, and Congress began to write federal legislation that established wildlife refuges and protected areas. Then by the 1970s the nation had become deeply concerned about the deterioration of the nation's surface waters, water pollution abatement became a federal mandate, and the first federal regulations over the destruction of wetlands were written. Later still the public called for federal protection of endangered species of plants and animals, most of which depend upon wetlands at some point in their life cycle. Finally, as the realization that all of these functions were tied in some way to the loss of our national wetlands, public policy became focused in the late 1980s on the single interest of wetland protection and preservation.

The piecemeal process by which the total body of wetland protection legislation developed, however, produced a fragmentation of programs among a set of federal agencies with very different mandates, philosophies, and spheres of activity. Before we examine the most important wetlands protection tool, the regulatory program authorized in Section 404 of the 1972 Federal Water Pollution Control Act Amendments (FWPCA), it is useful to look at some of the earlier developments.

THE EMERGING FEDERAL ROLE IN WETLANDS PROTECTION

The vast natural resources of the North American continent were prized by early settlers, in what was to become the United States, for their economic potential. The enormous energy of seventeenth and eighteenth and most of the nineteenth century expansion westward was directed toward harnessing these resources in the interests of first agricultural and then industrial development. In the context of that period, neither environmental degradation nor resource depletion was on anyone's mind. The earliest concern for the damage that economic development might cause to natural resources found expression in an awareness of the timber industry's impact on our rapidly disappearing forests, and it was through the new appreciation of the interrelationship between those forests and downstream water flows within the river basins that a sense of the hydrology of a watershed began to develop. A Wisconsin legislative commission pointed out the relationship between forest cover and streamflow in 1867; a New York state commission recognized the necessity of protecting the Adirondack mountain range for the benefit of the Hudson and other rivers in 1882; and in 1887, the American Forestry Congress urged the purchase of public lands at the source of streams for water supply benefits. In 1874, the American Association for the Advancement of Science directed Congress's attention to "the drying up of rivulets . . . and the growing tendency to flood and drought" (Brown and Murphy, 1955). The first forest preserves were set aside by Congress in 1891, and in 1897 the Organic Administration Act for national forests was passed, one of the main purposes of which was to secure the favorable conditions of water flow. The Forest Service was formally established in the USDA in 1902. In the interest of protecting our timber resources, watershed planning was gradually taking form, and it was doing so within the very federal agency that sponsored and encouraged the other agricultural program, drainage, that was disrupting the natural flows and destroying the wetlands.

Agricultural programs and policies historically have played an ambivalent role in the federal protection of wetlands, providing incentives to drainage on the one hand, and encouragement for the conservation of natural resources, including wetlands, on the other. The USDA, as the early advocate and benefactor of agriculture, encouraged and supported drainage along with other production-enhancement activities. It was particularly active in assisting with the rehabilitation of ditches in the 1930s, in response to the economic duress of the Depression. The Agricultural Conservation Program (ACP), which began in 1944, provided financial and technical assistance to both surface and subsurface drainage. Between 1942 and 1980, nearly 57 million acres of wet farmland (an unknown portion of which was actual wetlands) were drained with ACP funds, although as early as 1956 the USDA was attempting to curtail such assistance if new land was being brought into production. In 1962, PL 87-732 prohibited the USDA from providing any aid for wetland drainage in Minnesota and the Dakotas if it was determined by the Secretary of the Interior—on a farm by farm basis—that wildlife preservation would be harmed by the drainage (Office of Technology Assessment, 1984). Finally, directly after and in response to President Jimmy Carter's 1977 Executive Order 11990, requiring federal agencies to "take action to minimize the destruction, loss, or degradation of wetlands," the USDA officially stopped all cost sharing for wetland drainage. Yet remnants of the old program apparently lingered—in 1983, 37 farmers in 12 states were given federal help with drainage (Pavelis, 1987).

The role that agricultural subsidies have played in wetland destruction is less clear. These programs reduce the risk of farming by supporting commodity prices and, as such, have certainly encouraged the geographic expansion of agricultural production and thus the conversion of wetlands to cropland. One analyst looks upon them as having "the greatest marginal impact on wetlands of any federal subsidy program" (Scodari, 1997). Yet in the sense that their intent is to regulate and support the overall health of the agricultural industry, their impact on wetland conversion is secondary and viewed by others as having but "limited significance" (Office of Technology Assessment, 1984).

As the twentieth century progressed, both Congress and the agricultural community had become increasingly concerned with the negative impact farming activities could have on natural resources, particularly soil erosion. The USDA's Soil Conservation Service (SCS, recently renamed the Natural Resources Conservation Service or NRCS) was created in 1935 to put a halt to soil erosion. (The loss of topsoil had attracted the attention of Congress as early as 1928, when it passed the first federal appropriation for research on soil conservation.) The SCS was instructed to implement the Soil Conservation Act of 1937 and, in 1955, expanded its activities into flood control with the Watershed Protection and Flood Prevention Act, commonly known as the PL 566 programs. In 1970, when Congress passed the Water Bank Act that paid landowners not to drain wetlands, it read like a comprehensive wetlands protection bill:

> To preserve, restore, and improve the wetlands of the Nation, and thereby (1) conserve surface waters, (2) preserve and improve habitat for migratory water-

> fowl and other wildlife resources, (3) reduce runoff, soil, and wind erosion, (4) contribute to flood control, (5) contribute to improved water quality and reduce stream sedimentation, (6) contribute to improved subsurface moisture, (7) reduce acres of new land coming into production and to retire lands now in agricultural production, (8) enhance the natural beauty of the landscape, and (9) promote comprehensive and total water management planning.

In fact, the program was limited to Minnesota, North and South Dakota, and other states in the Mississippi River flyways of migratory waterfowl. And although its chief benefits flowed to wildlife habitat interests, its administration was assigned to the Agricultural Stabilization and Conservation Service (ASCS), now the Farm Services Agency (FSA), that agency within the USDA with the strongest allegiance to agricultural development programs.

The value of wetlands and the threat of drainage to wildlife survival was first appreciated in relation to migratory birds by the hunting community. It has been estimated that duck populations fell in the Prairie Pothole region of Minnesota and the Dakotas from a presettlement 15 million to 5 million in 1955, mainly owing to agricultural drainage (Schrader, 1955). The thrust of early wildlife habitat programs was the acquisition of lands to provide habitat for migratory birds, and funding came from a variety of sources. In 1929, the Migratory Bird Conservation Act established a Migratory Bird Conservation Commission. The Hunting Stamp Act of 1934 authorized the sale of "duck stamps" to raise funds for habitat purchases, and the Federal Aid in Wildlife Restoration Act in 1937 (Pittman-Robertson) gave grants to states for restoration projects for wildlife habitat. In 1950, the Federal Aid in Sport Fish Restoration Act (Dingell-Johnson) provided the same for fish.

Before 1960, the only habitat appreciated by Congress was for waterfowl and certain fish species (Committee on Environment and Public Works, 1982). After that date, interest in habitat protection shifted, gradually, from the objects of sport hunting and fishing to the broader wildlife population, for its own sake rather than the sake of the sportsmen. The Wetlands Loan Act of 1961 provided interest-free loans for wetland acquisition and easements. The Land and Water Conservation Fund Act of 1965 funded the purchase of natural areas, including wetlands, and has been used by the FWS to extend the National Wildlife Refuge System. In 1966, when the National Wildlife Refuge System Administration Act was passed, the heretofore unconsolidated system already included 472 national wildlife refuges covering 90 million acres, of which 35% was estimated to be wetlands.

The FWS itself was established within the Department of the Interior in 1956, by the Fish and Wildlife Act, and an important new role emerged for the new agency in 1958. The Fish and Wildlife Coordination Act, as amended in 1958, required that wildlife conservation be given consideration equal to the concern for other aspects in the water resource development projects of the Corps, Bureau of Reclamation, and other Federal Agencies, and gave the FWS and the National Marine Fisheries Service (NMFS), a Department of the Interior agency concerned with fish habitat, the authority to evaluate the impact on fish and wildlife of all new federal projects and federally permitted projects. In 1967 the FWS signed a Memorandum of Agreement (MOA) with the Corps, wherein the Secretary of the Army agreed that the Corps would "carefully evaluate" the advantages of any proposed activity that the Department of the Interior found would "unreasonably impair natural resources or the related environment." This was the beginning of the "public interest review" and established the role that wildlife habitat interests would play in the Section 404 Program—that of advisor. The importance of that advisory capacity, however, was not yet defined and would often be controversial.

WATER QUALITY AND SECTION 404 OF THE CLEAN WATER ACT

It was the environmental decade of the 1970s and the passage of the FWPCA of 1972 that created the single most important tool for the protection and preservation of wetlands. The decade had been inaugurated with the presidential signing, on January 1, 1970, of the National Environmental Policy Act (NEPA). Among the NEPA's stated purposes were to encourage harmony between people and the environment, prevent and eliminate damage to the environment, enrich the understanding of ecological systems and natural resources, and to establish a Council on Environmental Quality. The NEPA requires that all major federal actions must be evaluated in terms of their potential affect on the environment. The mechanism for doing this is, first, the preparation of an Environmental Assessment (EA) to determine whether a full Environmental Impact Statement (EIS) is necessary and, if so, the subsequent preparation of an EIS, a thorough and comprehensive description and evaluation of all impacts—direct, indirect, and cumulative—that might result from the proposed federal project.

The main intent of the FWPCA was to clean up the surface waters of the United States to the point where they were "fishable and swimmable," and to eliminate all pollutant discharges by 1985. The act's regulatory authority was assigned to the U.S. Environmental Protection Agency (USEPA), with a single exception. Section 404, the new device for wetlands protection, authorized the Secretary of the Army, acting through the Corps of Engineers (the Corps), to issue permits "for the discharge of dredged or fill material into the navigable waters at specified disposal sites."

The Corps appears, at first glance, to be the least likely federal agency to have responsibility for wetland protection. It was created in 1802 to "erect and maintain frontier forts and other defense facilities." By 1824, the Corps had been authorized to undertake river and harbor improvements promoting navigation, and in 1890 it was given authority to regulate all the construction activities and deposition of refuse into navigable waters. In 1899, Congress passed the precursor of today's 404 program, Section 10 of the Rivers and Harbors Act, which required that permits be obtained from the Corps for the construction of bridges, dams, or dikes across navigable waters of the United States. Section 10 operated under a definition of navigable waters related to the past, present, or future navigable capacity of waters below the mean high water line. Its responsibilities, therefore, were limited to obstructions of actual navigation.

Congress was slow to give the Corps any authorities that were not tied directly to navigation, and it was not until 1936 that national legislation even acknowledged flood control to be a responsibility of the federal government, to be carried out nationwide by the Corps of Engineers (Arnold, 1988). Large tracts of riverine wetlands were destroyed by the flood control activities of this branch of the Army subsequently, as thousands of miles of agricultural and urban levees were constructed to protect residential and commercial development and to allow the pumping, draining, and production of agricultural crops on riverine wetlands. Yet when Congress passed the FWPCA in 1972, it was this agency that was assigned responsibility for implementing the Section 404 permit process.

Because the Corps was the nation's largest navigational dredger and already responsible for issuing Section 10 permits for obstructions to navigation, some viewed the expansion of this role to include pollutant discharges as a logical extension of the agency's existing duties. Others explain the assignment of 404 responsibilities to the Corps more cynically (Blumm and Zaleha, 1989), as a move to prevent the USEPA from acquiring regulatory authority over Corps activities. As seen from the latter viewpoint, the Corps' role in the FWPCA was simply to provide them with an exception from the new regulatory program—the Corps' way of avoiding the permits that would otherwise be required by the USEPA under the Act's Section 402, the National Pollutant Discharge Emission System (NPDES). The actual assignment of 404 responsibilities to the Corps, whatever the reason for it, was to shift the goals of the Corps' regulatory authority from the protection of navigation to the improvement of surface water quality.

Although Section 404 of the FWPCA gave the implementation of the program to the Corps, it also, in Section 404(b), required that USEPA develop the guidelines for the program that would, in essence, set the environmental standards for the program. In Section 404(c), where the USEPA is authorized to prohibit, deny, or restrict the specification of a disposal site, thus being given a veto power over Corps activities, Congress enumerates the environmental damages it generally seeks to avoid by Section 404: "an unacceptable adverse effect on municipal water supplies, shellfish beds, and fishery areas (including spawning and breeding areas), wildlife, or recreational areas." The USEPA issued its 404(b) guidelines on September 5, 1975, developing the definition of "adverse impacts" and requiring permitted activities to avoid "discharges that disrupt aquatic food chains and destroy significant wetlands, avoiding degradation of water quality, and protecting fish and shellfish resources." The guidelines made it clear, as well, that permits should be issued only where the permitted activity is clearly water dependent, and where it can be demonstrated that there are no less environmentally damaging, practicable alternatives.

Although the explicit objective of the FWPCA was to "restore and maintain the chemical, physical, and biological integrity of the Nation's waters," the Corps shocked the environmental community by continuing to limit the jurisdiction of their permitting process to the traditional "navigable waters" under their 1899 Rivers and Harbors Act Section 10 authority. A series of court cases filed in 1974 and 1975 culminated in *NRDC* vs. *Callaway*, filed by the National Resource Defense Council and the National Wildlife Federation, and resolved in 1974 by a court order that the Corps expand its jurisdiction to conform with the broad intent of the 1972 law. Congress exploded. In the 93rd Congress, in session from 1973 to 1974, there had been little interest in the new water quality legislation, and most of the wetland-related legislation that was introduced (14 bills) aimed to establish and maintain wildlife refuges (Committee on Environment and Public Works, 1982). In the 1974–1975 session, however, 27 bills were introduced to amend Section 404, and the House attempted to restrict the 404 jurisdiction to "currently navigable" waters. The administration successfully opposed such a change, arguing that it would leave up to 85% of the nation's wetlands unregulated. The 95th Congress, in their 1977–1978 session, introduced 12 more bills to amend 404 and finally passed PL95-217, amendment to the FWPCA that henceforth became known as the Clean Water Act (CWA), without placing limits on the jurisdiction of Section 404 as defined by the court in 1974. Yet debate over the 404 jurisdiction was continued into the 96th and 97th Congresses by members intent upon further limiting its scope. In the 10-year period between 1973 and 1982, 48 bills had been introduced to amend Section 404 (Committee on Environment and Public Works, 1982).

Interestingly, while the 95th Congress was fighting over the CWA amendments, they were also passing two other wetland-related bills. In October, 1978, amendments to the 1934 Hunting Stamp Act, now called the Migratory Bird Hunting and Conservation Stamp Act (PL95-552), raised the price of stamps to bring more money into the federal coffers for wetland habitat. In the same month, the Fish and Wildlife Improvement Act (PL95-616) was passed, making administrative changes that facilitated the acquisition of migratory bird habitat by the FWS by amending several earlier laws. It was noted during the Congressional debates on these bills that annual acquisition of migratory bird habitat had declined from 159,000 acres in 1970 to 75,000 acres in 1977.

In 1976, Congress had amended and strengthened the Wetlands Loan Act, which was, at that time, the principal means of enabling the FWS to protect migratory birds and their habitat. Then, in 1980, the 96th Congress amended the Water Bank Program to make it more effective. And in 1986, the North American Waterfowl Management Plan resulted from an agreement between the United States and Canada to protect, enhance, restore, and create 6 million acres of wetlands for waterfowl habitat in North America. The plan consisted of joint ventures among public and private entities and was to be coordinated by the FWS. The North American Wetlands Conservation Act was passed subsequently to support the implementation of the plan. It extended the wetlands habitat protection aspects beyond those important only to waterfowl, included grants for wetlands conservation projects, and was administered by the FWS. In 1986, the Land and Water Conservation Fund Act was strengthened by amendments in the Emergency Wetlands Resources Act of 1986, which also required states to include acquisition of wetlands in the Comprehensive Outdoor Recreation Plans that they were required to prepare to be eligible for funding under the former Act. It was becoming clear that federal lawmakers felt far more comfortable providing wetland protection that would provide wildlife habitat, by federal acquisition, than they did regulating the uses of privately owned lands for water quality purposes, as they had done in Section 404.

Although the territorial jurisdiction of the 404 permitting process remained intact in the 1977 CWA, other features of the program did not (Congressional Research Service, 1978). A substantial expansion of the legislative language of Section 404 potentially relaxed the environmental protections of the program in three new ways:

- Section 404(e) allowed the Corps to issue general permits for categories of activities determined to have "minimal adverse environmental effects" either separately or cumulatively.
- Section 404(f) exempted from the permit process the discharge of any dredge or fill associated with normal farming, silviculture, or ranching activities and certain maintenance and construction activities.
- Several sections of the 1977 legislation also allowed and established conditions for the transfer of the 404 program to the states, under the purview of the USEPA.

On July 19, 1977, the Corps revised its regulations to include wetlands in the definition of the waters under its permitting jurisdiction, acknowledging that "wetlands are vital areas that constitute a valuable public resource, the unnecessary alteration or destruction of which should be discouraged as contrary to the public interest" and affirming that wetlands should not be filled for non-water dependent projects (as outlined in the 1975 USEPA guidelines). These Corps regulations proposed a review and consultation with the natural resource agencies—FWS, NMFS, SCS, USEPA, and state agencies—that could theoretically result in "elevation" of disputes among agencies to a higher level of Corps review, either to the division level or to the headquarters in Washington.

Then, in December of 1980, the USEPA issued revised guidelines emphasizing water dependency tests and presumption against alterations because, according to one observer, "some Corps districts were issuing Section 404 permits for non water-dependent activities when there were less environmentally damaging alternatives available" (Kruczynski, 1990). An advisory opinion issued the previous year by Attorney General Civeletti had established the

USEPA, rather than the Corps, as the ultimate interpreter of the scope of 404 jurisdiction (Blumm and Zaleha, 1989).

In 1981, the pendulum swung again, when newly elected President Ronald Reagan's Task Force on Regulatory Relief targeted Section 404, and in 1982 the Corps issued, as a "regulatory relief measure," new regulations that expanded the nationwide permit program, 404(e). It also renegotiated an MOA with the USEPA, FWS, and NMFS that limited the influence of those agencies on the 404 permitting process. Sixteen environmental organizations filed suit in December of 1982 (*National Wildlife Federation* vs. *Marsh*), claiming that the Corps had ignored the 404(b) guidelines, illegally expanded the use of nationwide permits, and reduced the ability of the resource agencies to impact the Corps decision making. The suit was settled in favor of the plaintiffs in February of 1984 and the Corps published revised regulations in October of 1984 and November of 1986, emphasizing the 404(b) guidelines. A new interagency MOA was negotiated in 1985 strengthening the role of the resource agencies in cases of dispute with Corps permit application approvals. In their 1987 amendments to the CWA, Congress made only minor changes to Section 404.

Section 404 had survived, but the jurisdictional issue that had consumed Congress in the 1970s was not to go away. A 1982 report prepared for the Senate Committee on Environment and Public Works (1982) concluded that "with public views ranging across the spectrum—from those who consider the expansive jurisdiction of 404 to be the most viable means of accomplishing wetland protection nationwide to those who consider the program overly burdensome, particularly to small landowners and developers regardless of congressional intention to safeguard wetlands—any upcoming debate will be lively."

FARM POLICY IN RESPONSE TO 404

Section 404, in fact, appeared to regulate only about 20% of the activities that destroy wetlands (U.S. General Accounting Office, 1991), because activities related to "silviculture, farming, and ranching" had been exempted by Section 404(f) of the CWA amendments of 1977. In 1985, Title XII, Subtitle C, of the Food Security Act (PL 99-198), popularly known as "Swampbuster," attempted to fill the void left by these exemptions made to the Section 404 program. This Swampbuster legislation did not prohibit the draining of wetlands, but it did discourage it by denying commodity payments to landowners for crops planted on wetlands that were converted after December 23, 1985. The Food, Agriculture, Conservation, and Trade Act of 1990 (the 1990 Farm Bill) went even further by withdrawing benefits from landowners who drained land, whether or not they actually planted on it.

A second program, the Conservation Reserve Program (CRP), was created by Title XII of the 1985 legislation and amended by the 1990 Farm Bill. The

CRP allowed the USDA to purchase 10- to 15-year easements on highly erodible lands, thus removing them temporarily from production. Cropped wetlands were eligible for enrollment in CRP from 1989 until the new Wetlands Reserve Program (WRP) program was created as part of the 1990 Farm Bill. The WRP authorized USDA to purchase 30-year or permanent easements for the protection and restoration of wetlands on lands that had been previously drained and cropped. The original program goal was to enroll up to 1 million acres by 1995, although by 1997 it had enrolled less than half of that amount.

Another USDA agency, the Farmers Home Administration (FmHA), which was a source of low cost loans to farmers, was given a role in wetland conservation in 1987, in the Agricultural Credit Act. This legislation established the FmHA Conservation Easement Program, under which properties that had defaulted and been repossessed by the FmHA could be placed in conservation easements. The 1990 Farm Bill reaffirmed this authorization and allowed for debt cancellation in return for acceptance of conservation easements. The FWS facilitated the process by providing review and recommendation of lands targeted for the program.

The nation's attention to coastal wetlands emerged gradually out of a general concern for the protection of coastal resources that developed in the 1970s, and the resulting legislation was housed in several different federal agencies. The Coastal Zone Management Act was first passed in 1972 and reauthorized in 1990. It encouraged the voluntary participation by coastal states in a program that coordinated federal resources, facilitated planning, and made grant money available for more intensive management of coastal resources that included, without targeting, large acreages of wetlands. The administration of the act was assigned jointly to the Office of Coastal Zone Management in the USEPA and to the National Oceanic and Atmospheric Administration (NOAA) in the Department of Commerce. The Coastal Barrier Resources Act of 1982, administered by the FWS's Division of Habitat Conservation and a stronger piece of legislation from a protectionist point of view, was passed to prohibit federal funds such as the national flood insurance program from going to development activities within designated units of the Coastal Barrier Resources System. The Coastal Barrier Improvement Act of 1990 expanded the jurisdiction of this system to include large acreages of wetlands. Finally, in 1990, Congress passed the Coastal Wetlands Planning, Protection, and Restoration Act (PL 101-624), which specifically targeted coastal wetlands and authorized the FWS to provide matching grants for the acquisition, management, restoration, or enhancement of coastal wetlands, particularly in Louisiana.

WETLANDS FOR THEIR OWN SAKE

One of the most notable changes in public thinking since the environmental decade of the 1970s has been the development of a view of wetlands as having an intrinsic worth that includes not only functional but also recreational, aesthetic, and even moral values. Hugh Prince has traced this environmental sensitivity back to Aldo Leopold's *A Sand Country Almanac*, published in 1949, which developed the ethic of wild lands and the necessity of living in respectful harmony with nature (Prince, 1997). Prince points out that the Sierra Club exploded from 15,000 mountaineers in 1960 to 100,000 environmentalists in 1971, 250,000 in 1981, and 500,000 in 1991. Several pieces of national legislation were passed in the 1960s that could be said to subjugate economic man to nature. The Wilderness Act of 1964 provided the first legislative guarantees of "freedom of human interference for wildlife and ecosystems" and the 1966 Endangered Species Act was replaced by a much stronger one in 1973. Congress passed the Wild and Scenic Rivers Act in 1968 to protect free flowing rivers from water resources development projects, mainly in the western states of Oregon, Alaska, and California.

Consistent with this focus on natural resources for their own sake, the Conservation Foundation, at the request of the USEPA, convened a National Wetlands Policy Forum in 1987. The final report of this body opened with, "The United States urgently needs a better system for protecting and managing its wetlands" (National Wetlands Policy Forum, 1988). It identified wetlands as providing public more than private benefits and claimed that the programs of the last two decades have "unfortunately . . . addressed only limited aspects of the wetlands protection problem and . . . have been adopted haphazardly and incoherently." The forum recommended the national wetlands policy statement that has subsequently driven our national thinking and talking about wetlands:

> . . . to achieve no overall net loss of the nation's remaining wetlands base, as defined by acreage and function, and to restore and create wetlands, where feasible, to increase the quality and quantity of the nation's wetlands resource base.

The forum also made recommendations in the broad categories of (1) protecting the resources; (2) improving the protection and management process; and (3) implementing the recommendations of the forum.

On February 9, 1989, President George Bush affirmed the "no net loss" goal, and in May of 1989, he established an interagency task force on wetlands under the Domestic Policy Working Groups to address the program requirements necessary to achieve no net loss. The goal statement was reaffirmed in a 1990 MOA signed by the USEPA and the Corps, and it has subsequently appeared in all significant official documents relating to the federal government's role in wetland protection. Such a statement was remarkable because, under existing programs and policy, overall net loss had been continuing, according to the most recent data, at a pace of more than 290,000 acres a year (Frayer et al., 1983).

The optimism reflected in the no net loss goal was made possible entirely by the emphasis being placed, as the 1990s opened, on compensatory restoration—a concept that is discussed in the following chapter. Although a new set of controversies have developed around the concept of compensatory restoration, most of them are derived from the same issues that have shadowed wetland protection since the 1972 passage of the regulatory program.

LINGERING WETLANDS ISSUES

Twenty-five years after the passage of the FWPCA and the Section 404 wetlands regulatory legislation, many of the early issues are still being debated, and both the proponents and opponents of wetlands protection remain discontent. On the one hand, landowners continue to be frustrated by the unreasonableness and unfairness of the 404 regulatory process, a frustration vented in arguments over jurisdictional and "takings" issues. Wetlands protectionists and the resource agencies, on the other hand, continue to be alarmed by the impact of general permitting and sequencing of the permitting review process that threaten to undermine the substantive goals of the CWA.

The jurisdictional argument, in the 1970s, had been over what geographic categories of waters were regulated under the 404 permitting process and who determined them. That issue was addressed and somewhat resolved on January 19, 1989, when the Corps and the USEPA signed an MOA allocating the jurisdictional decision-making to the Corps (reversing the earlier Civeletti decision) yet at the same time requiring the Corps to implement the USEPA guidance and report fully to the USEPA on any jurisdictional policy change and actions they might take. In the late 1980s and 1990s the jurisdictional arguments began to revolve around the question of what actually constituted a wetland; they were arguments about the scientific basis, or lack of it, for the definition itself. Wetlands are so saturated by water, over time, that changes in the soil and limitations of the plants occur, and the resulting hydric soils and hydrophytes (wetland plants) are used as evidence of the existence of wetlands where the actual presence of water may be missing at any given time. More than 2,000 soil types and more than 7,000 plant types have been associated with wetlands across the country. One critic of the regulatory program has claimed that proper identification of a wetland requires that a landowner be thoroughly conversant not only with a Corps 150-page guidance manual but also with two other ponderous publications: *Criteria for Hydric Soils* and *The National List of Plants that Occur in Wetlands* (Tolman, 1997). The identification of dry areas (at least in certain seasons) as wetlands and the assignment of rigid wetland boundaries to an observably undifferentiated landscape by the regulatory agencies were frequently perceived by the regulated community as arbitrary and capricious.

The Corps guidance manual identifying and delineating wetlands was issued in 1987. In 1989, the Corps, along with the USEPA, the FWS, and the SCS, issued a new manual that was immediately criticized as an attempt to expand the geographic jurisdiction of wetlands regulation. In May of 1990, the participating agencies initiated an evaluation of that manual and concluded that they needed to reexamine some key issues. Then in August of 1991, the USEPA proposed revisions of the 1989 document that resulted in a "virtually unprecedented 80,000 plus public comments" (Dennison, 1997) and they withdrew the 1989 manual altogether, reverting back to the 1987 document for wetland identification guidance.

A new definition of wetlands was adopted—locations typically inundated for 15 consecutive days during the growing season. It was not satisfactory to the regulated community, however, and the 104th Congress entertained but did not adopt proposals that the definition be loosened to apply only to properties inundated for 21 days, a change that would have exempted roughly 85% of the wetlands subjected at that time to Swampbuster rules (Heimlich et al., 1997).

In 1993, Congress had asked (through the USEPA) the National Research Council of the National Academy of Sciences to reexamine the definition issue. The resulting 1995 report (National Research Council, 1995) proposed that an entirely new manual be prepared, suggested a 14-day "provisional" duration threshold, and offered the following reference definition of wetlands:

> A wetlands is an ecosystem that depends on constant or recurrent, shallow inundation or saturation at or near the surface of the substrate. The minimum essential characteristics of a wetland are recurrent, sustained inundation or saturation at or near the surface and the presence of physical, chemical, and biological features reflective of recurrent, sustained inundation or saturation. Common diagnostic features of wetlands are hydric soils and hydrophytic vegetation. These features will be present except where specific physicochemical, biotic, or anthropogenic factors have removed them or prevented their development.

Regardless of the definition of a wetland, there still existed a strong objection to the 404 regulatory concept—that a private property owner could be deprived of the right to use that property as he or she chose. The "takings" issue, derived from language in the fifth amendment of the U.S. Constitution ("nor shall private property be taken for public use without just compensation"), had been given some support from court cases (Blumm and Zaleha, 1989) and was argued even more strongly in the 1990s than it had been in the 1970s. In 1988, in response to the "takings" cases of the Supreme Court's 1987 term, President Reagan attempted to issue an executive order requiring that a Takings Implication Assessment (TIA) be prepared and submitted to the Office of Management and Budget for any proposed government actions that would affect private property. The 104th Congress proposed and debated, but failed to pass, legislation that would have required the federal government to compensate landowners when regulations diminished the value of a portion of a property by 20–30%, a bill that would have affected, at a minimum, the Endangered Species, Section 404, and the Swampbuster provisions of the Farm Bill.

The proponents of strong wetlands protection, on the other side of the issue, were criticizing the general permits authorized by Section 404(e), Nationwide

Permit (NWP) 26 in particular, as compromising the substantive intent and purpose of Section 404. NWP 26 relates to *Activities in Isolated Waters and Headwaters* and, according to the critics, affects at least 25,000 acres annually and is "the single largest source of wetlands loss in America" (Caputo, 1997). The nationwide permits overall, although they were limited to 5 years' duration, have been charged with exempting 17 million acres of wetlands from regulation (Crane, 1997) even though the 1984 *National Wildlife Federation* vs. *Marsh* settlement had required the reinstatement of a 10-acre limit for such permits and thus reduced their scope. The Corps, after reviewing more than 4,000 comments on the renewal of NWP 26 in 1997, agreed to phase it out by the end of 1998 and, instead, introduced a new set of NWPs in the summer of 1998 that environmentalists argue will open the door for increased wetlands destruction.

Finally, the advisory nature of the Section 404 public review and its impact on the decision-making process itself have been the continuing object of criticism. Blumm and Zaleha (1989) have pointed out that the Corps issues most of the permits that are applied for, citing a study done in Louisiana between 1980 and 1986 where 99.36% of all permit applications were approved. An area of particular concern has been the small role played by the agencies with the most wetland expertise—the FWS, the NMFS, and state natural resources agencies—and the failure of the USEPA to make use of their veto power, 404(c), having invoked it only five times in the first 15 years. Enforcement of the program is perceived as poor because the enforcement responsibilities are divided between the USEPA and the Corps. The General Accounting Office reported, in 1988, that "neither the USEPA nor the Corps has implemented an effective program of detecting violations" (Blumm and Zaleha, 1989).

The internal Corps review has also been criticized in the context of arguments about the proper sequencing of the evaluation criteria. The natural resource agencies have always emphasized that every attempt should be made to achieve first *avoidance*, and then *minimization* of environmental impacts; only after that should *compensatory restoration* be considered, but that none of these should be considered at all until the applicant has first demonstrated that the proposed activity is water dependent and that no reasonable and less damaging alternatives exist. Critics have perceived the Corps review process as giving too little attention to the early steps, and moving too rapidly to the later ones. Rigorous adherence to this sequencing is reportedly not only rare but has even been argued to be inappropriate (Clark, 1990).

THE CLINTON ADMINISTRATION'S WETLANDS PLAN

On August 23, 1993, the White House Office on Environmental Policy issued the Clinton Administration's Wetlands Plan—"Protecting America's Wetlands: A Fair, Flexible, and Effective Approach."

The plan explicitly is derived from five principles:

1. Programs must support the short-term goal of no overall net loss and the long-term goal of increasing the quality and quantity of the nation's wetlands resource base.
2. Regulatory programs must be efficient, fair, flexible, and predictable.
3. Nonregulatory programs must be encouraged to reduce the federal government's reliance upon regulatory programs.
4. The federal government should expand partnerships with states.
5. Federal wetlands policy should be based upon the best scientific information available.

These principles reflect the problems confronting the wetlands policy in the 1990s. The federal agencies that were familiar with our wetlands policies and their effects recognized that the country was continuing to suffer net losses, that many in the regulatory community considered the wetland programs to be inefficient, unfair, and inflexible, that there wasn't enough reliance upon nonregulatory programs, that the federal government wasn't giving the states a large enough role in wetlands regulation, and that federal wetlands policy was not well grounded in "good science."

There was something in the Clinton plan for everyone concerned about our national wetlands policies. The reforms it promised included the following:

1. Affirmation of the no net loss goal.
2. Setting up a 404 administrative appeals process for landowners.
3. Speeding up the 404 process.
4. Removing drained cropland from 404 regulatory jurisdiction.
5. Making the SCS the lead agency for identifying wetlands on agricultural lands.
6. Endorsing the use of mitigation banks.
7. Increasing funding for the USDA's Wetlands Reserve Program.
8. Providing incentives for watershed planning by states and localities.
9. Promoting wetland restoration through voluntary nonregulatory programs.

The plan discusses a wide range of issues related to national wetland policy, giving considerable attention to restoration, mitigation, and even mitigation banks. It was becoming clear to everyone involved, at this juncture, that the national wetland goals would never be achieved without increased reliance on wetland restoration. Better known as mitigation, it alone provided the mechanism for accomplishing no net wetlands loss in the short term and net gain in both quantity and quality in the long term.

REFERENCES

Arnold, J. L., *The Evolution of the 1936 Flood Control Act*, U.S. Army Corps of Engineers, Fort Belvoir, VA, 1988.

Blumm, M. C. and D. B. Zaleha, "Federal Wetlands Protection Under the Clean Water Act: Regulatory Ambivalence, Intergovernmental Tension, and a Call for Reform," *University of Colorado Law Review*, Volume 60, Number 4, 1989.

Brown, C. B. and W. T. Murphy, "Conservation Begins on the Watersheds," *Water, Yearbook of Agriculture*, U.S. Government Printing Office, Washington, DC, 1955.

Caputo, D., "Wetlands Loss Study Out," Cleanwater-Info@igc.org., September 17, 1997.

Clark, J. R., "Regional Aspects of Wetlands Restoration and Enhancement in the Urban Waterfront Environment," *Wetland Creation and Restoration: The Status of the Science*, Volume II, Island Press, Washington, DC, 1990.

Committee on Environment and Public Works, U.S. Senate, *Wetland Management*, U.S. Government Printing Office, Washington, DC, 1982.

Congressional Research Service of the Library of Congress, *A Legislative History of the Clean Water Act of 1977*, Volume 3, U.S. Government Printing Office, Washington, DC, 1978.

Crane, S., "Army Corps of Engineers Issues Final Nationwide Permits," *Wetland Journal*, Winter, 1997.

Dennison, M. S., *Wetland Mitigation: Mitigation Banking and Other Strategies for Development and Compliance*, Government Institutes, Rockville, MD, 1997.

Frayer, W. E., T. J. Monahan, D. C. Bowden, and F. A. Graybill, *Status and Trends of Wetlands and Deepwater Habitats in the Coterminous United States, 1950s to 1970s*, Colorado State University, Fort Collins, CO, 1983.

Heimlich, R. E., K. D. Wiebe, R. Claassen, and R. M. House, "Recent Evolution of Environmental Policy: Lessons From Wetlands," *Journal of Soil and Water Conservation*, May–June, 1997.

Kruczynski, W., "Wetlands and the Section 404 Program: A Perspective," *Wetland Creation and Restoration: The Status of the Science, Volume II*, Island Press, Washington, DC, 1990.

National Research Council "Wetlands Characteristics and Boundaries," National Academy Press, Washington, DC, 1995.

National Wetlands Policy Forum, *Protecting America's Wetlands: An Action Agenda. The Final Report of the National Wetlands Policy Forum*, The Conservation Foundation, Washington, DC, 1988.

Office of Technology Assessment, U.S. Congress, *Wetlands: Their Use and Regulation*, U.S. Government Printing Office, Washington, DC, March, 1984.

Pavelis, G. A., "Economic Survey of Farm Drainage," *Farm Drainage in the United States: History, Status and Prospects*, U.S. Department of Agriculture, Washington, DC, 1987.

Prince, H., *Wetlands of the American Midwest*, University of Chicago Press, Chicago, 1997.

Schrader, T. A., "Waterfowl and the Potholes of the North Central States," U.S. Department of Agriculture, *Water: The Yearbook of Agriculture, 1955*, U.S. Government Printing Office, Washington, DC.

Scodari, P. F., *Measuring the Benefits of Federal Wetland Programs*, Environmental Law Institute, Washington, DC, 1997.

Tolman, J., ''How We Achieved No Net Loss,'' *National Wetlands Newsletter*, Volume 19, Number 4, Washington, DC, 1997.

U.S. General Accounting Office, *Wetlands Overview: Federal and State Policies, Legislation and Programs*, U.S. Government Printing Office, Washington, DC, November, 1991.

CHAPTER 5

WETLAND RESTORATION: COMPENSATION FOR LOSSES

By the time the National Wetlands Forum was convened in the summer of 1987, there was concern that wetland losses were continuing, in spite of 15 years of regulatory protection under Section 404 of the Federal Water Pollution Control Act (FWPCA) of 1972. In adopting their interim and long-term goals, the forum relied upon the execution of a strategy that had not been officially recognized in either the language of the act or in the Section 404(b) guidelines that defined its intent: wetland mitigation. Mitigation is the restoration, creation, or enhancement of wetlands to compensate for wetlands losses (Lewis, 1989) that, by the end of the 1980s, was being commonly used by applicants for 404 permits to compensate for their destruction of existing wetlands. Without such mitigation most wetland conservationists assumed that it would be impossible to achieve either the forum's interim or long-term goals, because permits were being issued regularly to allow the destruction of existing wetlands. Although the only wetlands loss data available to the members of the forum were for the period between the mid-1950s and mid-1970s, data forthcoming in the following year, reflecting the wetland losses between 1974 and 1983, proved this assumption to be correct.

STATUS AND TRENDS OF WETLAND LOSSES

In 1982, the FWS had completed a National Wetland Trends Study (Frayer, 1983) that estimated the existence of 99 million acres of wetlands in the mid-1970s, reflecting a loss of 11 million acres of wetlands since the mid-1950s.

Subsequently, the FWS made corrections to that study, concluding that 105.9 (rather than 99) million acres of wetlands existed in 1974. Using this new number, then, they calculated that an average of 458,000 acres had been lost annually between the mid-1950s and the mid-1970s.

The Emergency Wetlands Resources Act of 1986, requiring the FWS to update that inventory every 10 years, resulted in another assessment of wetland losses, published in 1991, that showed a further reduction of national wetlands to 103.3 million acres by the mid-1980s. This 2.6 million acre loss between 1974 and 1983 represented an average annual loss of 290,000 acres of wetlands. In the decade following the passage of Section 404, therefore, although average annual wetland losses had been reduced by 37% (from 458,000 to 290,000 acres) they still remained substantial. Examination of the different types of wetlands lost showed actual increases in a few categories, whereas palustrine forested wetlands, for example, suffered a loss of 3.4 million acres during that time period.

In response to a congressional mandate in the 1989 North American Wetlands Conservation Act, the FWS calculated and published, in 1990, an estimate of the total wetlands in presettlement America: 221 million acres in the contiguous United States. Of those, 53% had therefore been lost (leaving only 103.3 million acres) by the mid-1980s (Dahl and Johnson, 1991). Finally, in 1997, the FWS updated their status and trends reports, calculating an average of 117,000 acres of wetlands lost annually between 1985 and 1995. Average annual wetland losses today, therefore, are 25.5% of what they were prior to passage of the FWPCA and implementation of Section 404.

A second federal wetlands inventory is extrapolated from the Natural Resources Inventory (NRI) done by the Natural Resources Conservation Service (NRCS) in the USDA in 1982, 1987, and 1992; it uses a different methodology from that employed by the FWS but produces only slightly different results. The NRI estimated, for example, that between 1982 and 1992 we lost an annual average of 156,000 acres of wetlands.

The conclusion that wetland losses are continuing at all was challenged by one analyst (Tolman, 1997) who argued that wetland restoration efforts, mainly in volunteer programs, more than equaled the losses calculated by the NRI, and that there had actually been a net gain in wetland acreages nationwide, in 1995, of 69,000 acres. Although critics of this analysis (Heimlich et al., 1997) have acknowledged that the nation might, at last, be reaching the interim goal of no net loss, they and others have generally agreed that restoration statistics are at best incomparable to loss statistics and at worst totally misleading as representations of new wetlands created. The statistics that are generated by the voluntary restoration programs such as the Wetland Reserve Program (WRP) and Partners for Wildlife contain acreages that are only wetland enhancements or improvements and sometimes even non-wetland areas, and these numbers cannot be removed from the data in order to calculate the actual acres of new, restored wetlands. Further, there is no commitment by these participants in voluntary programs to keep the lands indefinitely enrolled in the programs, giving them only a temporary status as wetlands. The WRP, for example, now requires the NRCS to enroll one-third of the projects under annual contracts, one-third under 30-year easements, and only one-third under perpetual easements.

It has become clear, at this juncture, that the country is still experiencing some wetland losses, but the rate of these losses has diminished dramatically since Section 404 of the FWPCA went into effect. By using compensatory restoration, particularly with ratios greater than 1:1 (restored:destroyed wetlands), there is the opportunity to achieve the forum's goal of no net loss in the near future, and even to meet the National Research Council's goal of an overall gain of 10 million acres of wetlands by 2010. That accomplishment will be meaningless, however, if compensatory restoration does not provide us with functional equivalents of the lost wetlands. Although the language of public mitigation policy promises that today, it is a relatively new policy and represents a shift in emphasis that was not explicitly anticipated by the USEPA and the Corps in the early days of Section 404.

THE ORIGINS OF MITIGATION AS PUBLIC POLICY

Although mitigation is not mentioned in the 1972 FWPCA, the concept gradually achieved recognition in FWS programs, Corps projects, and USDA programs. The USEPA sponsored a report, at the end of the 1980s, that described the considerable advances that had been made in the science of compensatory restoration. Finally, in a 1990 MOA, it became an official part of the 404 permit process.

Early on, mitigation was mentioned in the Fish and Wildlife Coordination Act, as amended in 1958, and in the NEPA of 1969. The Council on Environmental Quality (CEQ), in 1978 regulations issued for the implementation of NEPA, defined mitigation to include the following:

- Avoiding environmental impacts altogether by not taking an action (or part of an action) that might lead to environmental degradation.
- Minimizing impacts by limiting the degree or magnitude of the action and its implementation.
- Rectifying the impact by repairing, rehabilitating, or restoring the affected environment.
- Reducing or eliminating the impact over time by preservation and maintenance operations during the life of the action.
- Compensating for the impact by replacing or providing substitute resources or environments.

The last of these five activities, compensatory restoration, is the de facto definition of wetland mitigation today and the one adopted in these pages.

Mitigation has, in fact, been used in the implementation of Section 404 since its inception. The FWS and the National Marine Fisheries Service (NMFS), two Department of the Interior agencies with wildlife and fisheries habitat-protection missions, were given very little official influence on the 404 permitting process, prompting them, in the view of one observer, to rely heavily upon mitigation to accomplish their own ends (Kryzinski, 1989). Those agencies were able to achieve compensatory replacement for wetland losses, in the process observed by William Kryzinski (as he implemented the program in USEPA's Region IV), where they could not otherwise influence the Corps' issuance of a permit to which they objected. In 1981, the FWS promulgated a policy stating that mitigation can be considered for proposals that

- Are ecologically sound.
- Select the least environmentally damaging alternative.
- Avoid or minimize loss of fish and wildlife resources.
- Adopt all measures to compensate for unavoidable loss.
- Demonstrate a public need and are clearly water dependent.

Compensatory restoration was attractive both to the reviewing agencies and to the permit applicants because it vastly facilitated the onerous and time-consuming 404 review process, but it was acceptable to the wetland protectionist community only to the extent to which it could effectively and predictably replace the lost wetland functions. Attempts to restore and create wetlands during the 1970s provided a body of evidence to support such effectiveness and predictability. In 1973, Congress had authorized the Corps' Dredged Material Research Program, which assessed the feasibility of developing habitats on dredged materials substrate. Participants in an annual "Conference on the Restoration and Creation of Wetlands" had been demonstrating increasing self-confidence in their craft since their beginnings as the "Conference on the Restoration of Coastal Vegetation in Florida," in 1974.

In 1989, the USEPA sponsored an examination of "Wetland Creation and Restoration: The Status of the Science," published in two volumes. In the executive summary, John Kusler and Mary Kentula listed a number of points about the status of wetland restoration. Selected points are paraphrased below:

- While hundreds of coastal estuarine mitigation projects have been constructed on the eastern seaboard, there are far fewer on the Gulf and Pacific coasts and even less has been done to restore inland wetlands. The most common and best down types of inland restorations are impoundments to create waterfowl and wildlife marshes, and the creation of marshes on dredged soil along rivers as part of the Corps of Engineers Dredged Material Program.
- It has been difficult to measure success because there are often no goals identified, and there is very little monitoring, either short- or long-term, to determine the outcome of the restoration activity.

- It appears to be easiest to restore estuarine, then coastal, and thirdly, freshwater marshes. It is more difficult to restore isolated marshes supplied by surface water, even more difficult to restore forested wetlands and almost impossible to restore isolated freshwater wetlands supplied by groundwater.
- In terms of the restoration of wetland functions, it is easiest to restore flood storage and conveyance functions; we're successful in restoring waterfowl production because we know so much about it; wetland aesthetics are easy for the visual effects and difficult in reproducing subtle ecological functions. There is substantial variation in our restoration of fisheries and food chain functions, depending on the specifics. Pollution control is relatively easy in removing sediments and more difficult in reducing toxics. As for groundwater recharge, we can't even assess it, much less recreate it.
- Short-term successes are deceptive because vegetation may be difficult to sustain, long-term fluctuations in hydrology are common, excessive sediment builds up and erosional equilibrium is tricky. Restorations often need midcourse corrections, and long-term management needs include water control structures, replanting, regrading, buffers, barriers, controls on pollution and invasion, and periodic dredging.

With the growing respectability of wetland creation and restoration as a science, therefore, the federal government was able to sustain its wetlands protection programs by shifting from a reliance on prohibition and minimization of wetland destruction to the encouragement of compensatory restoration that made these programs more palatable to the private property owner. Compensatory restoration was given an important role in the 1986 Water Resources Development Act (WRDA). Although the overall act continued the construction orientation of past WRDAs by authorizing 270 new projects, the potential for environmental damage from these works was reduced by the inclusion of provisions that, for the first time, required the Corps to develop a mitigation plan for every new project that they proposed for construction. The likelihood of the actual construction of these newly authorized projects was also reduced by another provision in the same act—an increase in the required local cost share. The 1990 WRDA went a step further, identifying "environmental protection" as a primary goal of water resources projects and explicitly affirming the forum's two national wetlands goals: no net loss in the short term and net gain in the long term.

Agricultural policy followed suit by explicitly referencing mitigation activities. The Farm Bills of 1985 and 1990 had included Swampbuster provisions to discourage new conversions of wetlands to cropland, and the Farm Bill of 1996 (the Federal Agricultural Implementation and Reform Act, or FAIR) gave persons who converted wetlands a greater flexibility to mitigate the loss of wetland functions and values through restoration, enhancement, or creation of wetlands. The FAIR even authorized a pilot program for agricultural wetland mitigation banking and allowed the USDA to waive the Swampbuster penalties if it was believed that a violation of wetland provisions was unintended.

The role of mitigation in satisfying the requirements of the 404 permit process was finally addressed formally in a series of memoranda in the early 1990s. On February 6, 1990, the USEPA and the Corps issued the first of these, an MOA concerning the "Determination of Mitigation Under the Clean Water Act Section 404(b)(1) guidelines," applicable to individual permits. The memo listed the activities included in the CEQ definition of mitigation and suggested that they can be reduced to three categories: avoidance, minimization, and compensation. It affirmed the goal of "no overall net loss to wetlands," and pointed out that this goal "may not be achieved in each and every permit action" (Paragraph IIB). In other words, without compensatory restoration, there cannot ever be no net loss of wetlands. The 1990 MOA addressed the controversial "sequencing" question by explaining that the Corps "first makes a determination that potential impacts have been avoided to the maximum extent practicable; remaining unavoidable impacts will then be mitigated to the extent appropriate and practicable by requiring steps to minimize impacts and, finally, compensate for aquatic resource values" (IIC).

The memo goes on to discuss first avoidance (IIC1), then minimization (IIC2), and finally, compensatory mitigation (IIC3) "required for unavoidable adverse impacts which remain after all appropriate and practicable minimization has been required." This latter section states that on-site is more desirable than off-site mitigation, that functional values lost should be considered in determining the compensatory restoration, that in-kind is preferable to out-of-kind, and that wetland restoration should be selected over wetland creation, in that "there is continued uncertainty regarding the success of wetland creation" (IIC3).

There is a brief discussion of mitigation banking as "an acceptable form of compensatory restoration under specific criteria" (IIC3), a promise that additional guidance on mitigation will be provided, and an admonition that "preservation" or purchases "may in only exceptional circumstances be accepted as compensatory mitigation" (IIC3). In paragraph IIIB the MOA says, "The objective of mitigation for unavoidable impacts is to offset environmental losses," and that it should provide "at a minimum, one to one functional replacement, with an adequate margin of safety to reflect the expected degree of success associated with the mitigation plan," adding that "this ratio may be greater where the functional values of the area being impacted are demonstrably high and the replacement wetlands are of lower functional value, or the likelihood of success of the mitigation project is low." Ratios of one to one for compensatory restoration are rarely used today; more typical ratios are 1.5:1 for restoration, 2:1 for creation, and 3:1 for enhancement.

What the 1990 MOA calls "compensatory mitigation" is defined as the "restoration of existing degraded wetlands or creation of man-made wetlands" but it also includes, in practice, enhancement, which upgrades the functions of all or part of an existing wetland. The "preservation" referenced in the 1990

MOA (sometimes referred to as "exchange") allows a 404 permit applicant to purchase or provide the money for the purchase of valuable wetland property that ensures its long-term protection, in exchange for the destruction of less valuable wetlands.

The 1990 MOA talks a lot about the replacement of functional values of wetlands, a condition that can be met only if there is a satisfactory analysis of the functional values of both the lost and the replacement wetlands. Several methodologies have been developed to assess the functional value of wetlands, including the Habitat Evaluation Procedure (HEP) developed by the FWS and the Wetlands Evaluation Technique (WET) developed by the Corps and Federal Highways (FHWA) in the U.S. Department of Transportation. HEP has been in use by FWS since the mid-1970s and was revised in 1980. It requires sophisticated analysis and formal training, the calculation of a Habitat Suitability Index (HSI) for each indicator species, and the measurement of site values and mitigation "credits" in terms of Habitat Units (HUs). The first version of WET, WET 1.0, was developed in 1983 and modified (WET 2.0) in 1987, giving ratings for wetland functions in categories of effectiveness, opportunity, and social significance. Another analytic tool, the Hydrogeomorphic Classification System (HGM) became popular in the 1990s, and is specifically recommended in the Clinton Administration's Wetlands Plan for use in assessment of wetland functions. This system classifies wetlands based upon characteristics such as topographical position in the landscape, the source of water, and the direction of water within the wetland itself. The major classes are riverine, depressional, slope, flat, estuarine fringe, and lacustrine fringe. Subclasses for each of these are defined regionally. The use of these techniques has proved to be awkward and, as yet, has not been widespread.

No one knows quite what the costs of compensatory restoration are, nor even how to measure them. Private costs, calculated as the costs imposed upon a developer by the 404 regulatory process, have recently been estimated to be between $40,000 and $115,000 per acre (Vanderpool, 1998). The social welfare perspective would count compensation payments from the government to the landowner as transfer payments and only count the cost of administration as a cost of the 404 and Swampbuster programs. Administrative costs of the 404 permitting program have been calculated to be $78 million in the single year of 1995 but there is no way to calculate the benefits, which would be the number of acres of wetland losses prevented in that time frame. The WRP program, on the other hand, has actually been able to estimate the per acre administrative cost of the program: $70 per acre restored with an additional $5 per acre to monitor each acre over its life.

MITIGATION BANKING

The Clinton Administration's 1993 Wetlands Plan gave a strong endorsement to compensatory restoration and an official White House nod to the relatively new concept of mitigation banking. Mitigation banking creates off-site mitigation "credits" that can be purchased by permit applicants in lieu of implementing mitigation actions on their development sites. It takes the concept of compensatory restoration and moves it a step further away from the original Section 404 objective of preserving and protecting existing natural wetlands.

The Clinton plan acknowledged that "restoring some former wetlands, that have been drained previously or otherwise destroyed, to functioning wetlands is key to achieving the administration's interim . . . and its long-term goal" and among its 12 major initiatives was an endorsement of the use of mitigation banks in order to "increase the predictability and environmental effectiveness of the Clean Water regulatory program." It described mitigation banking as the fourth step in the 404 review sequence, following avoidance, minimization, and compensation when they are not appropriate, practicable, or as environmentally beneficial as the development site.

The FWS had developed the concept of mitigation banking in the early 1980s, and an MOA authored by the Corps and the USEPA the day before the release of the Clinton Wetlands Plan addressed the applicability of mitigation banking to the 404 process (Dennison, 1997). The Corps initiated the National Wetlands Mitigation Banking Study, and further elaboration and extension of the mitigation banking concept to a full range of federal activities was provided in the public register notice of November 28, 1995. These regulations identified the Corps as the lead agency for mitigation banking with the exception of those banks proposed solely to comply with Swampbuster, which would be reviewed by the NRCS. The Federal Register notice set the goal of providing economically efficient and flexible mitigation opportunities while at the same time including "the need to replace essential aquatic functions which are anticipated to be lost through authorized activities within the bank's service areas." In selecting the sites for mitigation banks, agencies were asked to ensure that they possess the "physical, chemical, and biological characteristics to support establishment of the desired aquatic resources and functions." "Compatibility with adjacent land uses and watershed management plans" were identified as "important factors for consideration."

Mitigation banks were at first developed exclusively by single-user public entities to offset their own mitigation requirements, but gradually, private (non-applicant) entrepreneurs began to develop what are seen as market-based banks. Proponents argue that banks will reduce and resolve the problems associated with project-specific compensatory restoration and that they provide an excellent vehicle for incorporating mitigation into watershed planning. Opponents fear that banks will be used even more than are project-specific off-site mitigations as easy and inadequate alternatives to the hard work of avoidance and minimization of impacts on wetlands. By early 1998, there were close to 200 approved wetland mitigation banks in existence.

THE LIMITATIONS OF MITIGATION

Through the use of mitigation and mitigation banking, the implementation of our wetland protection programs has come to rely heavily upon the ability of wetland creation, restoration, and enhancement techniques to compensate fully for the lost wetland functions. Scientists and regulators working in the field continue to express strong reservations, however, about the effectiveness of these techniques. The difficulties seem to fall into the following categories:

- *Technical impossibility of compensating for the functions of certain rare and complex wetland types* such as oligotrophic bogs or those supporting very old forests. The scientific community agrees that certain wetland types and functions are irreplaceable; there is far less agreement upon which ones and where these are, however, or how these questions are even answered.
- *Failure to assess adequately the functional attributes of the lost wetlands, through either technical inability or operational failure.* The role and values of the affected wetlands are not fully appreciated, in many cases, and even when they are, the regulatory process is not geared to take them into account.
- *Failure to replace adequately the functional attributes of the lost wetlands, through either technical inability or operational failure.* Leonard Shabman and Paul Scodari (1995) have summarized the complaints of many workers in the field in their conclusions that much promised compensatory restoration is never done, that when it is done it is done badly, that when it's done well no one monitors it, that when it is monitored and determined to be deficient there is no one responsible for rectifying problems, and finally, that even if it is determined to be successful, there is no long-term maintenance that ensures that it remains successful. Other studies supported these conclusions. Robin Lewis reported that in 1984 only 4.5% of 174 Florida restoration projects reviewed met the permit success criteria. He reported later that only 50% of these "paper mitigation" projects had even been started (Lewis, 1997). The New England District Corps office reported in 1989 that 36% of 100 projects reviewed had failed.
- *Failure to identify appropriately and reproduce the value of lost wetland functions within their temporal and geographical context.* The value of any given wetland depends upon its relationship to the land and water regimes and ecosystems of which it is a part. Yet there has been no mechanism for assessing this contextual value until the recent attention to the role of watershed planning in wetland mitigation.

WETLANDS MITIGATION IN THE CONTEXT OF WATERSHED PLANNING

Concurrent with the developing national interest in wetland restoration has been a growing emphasis within the resource agencies, and the USEPA in particular, on making water resources management decisions within the context of the total needs of individual drainage basins. This watershed planning approach, if done well, promises to solve the major difficulties encountered in compensatory restoration. Mitigation banks, in fact, have been seen as the very mechanism by which wetland restoration will be successfully incorporated into such plans. Although the proper and successful implementation of watershed planning requires a level of technical expertise and bureaucratic sophistication that is probably still beyond our reach, its achievement may allow us to achieve *genuine* "no net loss" and perhaps even long-term gains in wetland values and functions.

Different federal agencies and programs have acknowledged the value of planning in the broadest possible context, by different routes. The USEPA has established the watershed as the geographic unit within which to control nonpoint source pollution and, in 1991, published *The Watershed Protection Approach*; in February of 1995, it produced a Fact Sheet on Wetlands and Watersheds (U.S. Environmental Protection Agency, 1995) that suggests looking at the whole system—land, air, and water—to develop management plans for aquatic resources; and in 1996, it published a Watershed Approach Framework. In 1993, the White House Office on Environmental Policy established an Interagency Ecosystem Management Task Force and the four agencies who are responsible for federal ownership of about 30% of our total national land surface area—the National Park Service, the Bureau of Land Management, the FWS, and the USDA's Forest Service—are developing ecosystem approaches to their land and natural resources management (U.S. General Accounting Office, 1994). The NRCS in the USDA has established ecosystem-based management of natural resources. Local planning bodies have adopted subwatershed and regional approaches. The Corps' Special Area Management Plans (SAMPS) and the USEPA's Advanced Identification (ADID) process share some of the attributes of watershed planning, as well. These are all promising first steps in the right direction, but it would be inappropriately optimistic to call them anything more than first steps.

Watershed planning sometimes has been single purpose and has not always included wetland preservation and restoration components. Good watershed approaches to environmental planning and management, according to the USEPA, contain strong partnerships with stakeholders; a geographic focus and management techniques based on strong science and data; and coordination of a wide range of programs including drinking water source protection, waste management, point and nonpoint source pollution, air pollution, pesticide management, and wetlands protection.

In the introduction to a 1995 publication containing a wide range of papers on wetlands and watershed management (Kusler, 1995), principal editor Jon Kusler has made it clear that the most important feature of a watershed planning context is the inclusiveness of the water and landscape regimes. In that context, Kusler says, the role of wetlands in relation to water quality, flood storage and conveyance, wildlife and fisheries habitat, and recreation can be assessed. Some of the problems currently plaguing wetland protection programs that Kusler believes can be resolved by adopting the watershed approach include the following:

- Unsuccessful restorations because of lack of control over outside influences on water quantity and quality.
- Failure to identify wetlands functions in relation to other parts of the landscape and water regime.
- Failure to recognize cumulative negative impacts on wetlands.
- Unresolvable conflicts with other water resources management objectives and programs.
- Inability to identify optimum restoration sites.
- Inability to identify wetland functions.
- Hydrologically failed projects.

Good watershed-based planning, according to Kusler, includes the following components:

- Identification of specific problems and needs.
- Involvement of all the key players.
- Scientific understanding of the key relationships between wetlands and water regimes.
- Specific wetland/ecosystem goals.
- Good mapping.
- Analysis of the relationship between wetlands and other elements.
- Major involvement by locals.
- Consideration of implementation techniques "up front."
- Involvement of all—not just wetlands—water resources managers.

Although both the USEPA's and Kusler's criteria for successful watershed-based programs are well beyond our present institutional capabilities, many states and regional agencies are taking the first steps toward their accomplishment. In Massachusetts, the Executive Office of Environmental Affairs established a Wetlands Restoration and Banking Program in March of 1994 to tie compensatory restoration and mitigation banks to overall watershed deficits and needs. Twenty-eight watersheds have been identified, within which they are first identifying restoration sites, second, identifying watershed needs and goals, and third, screening the sites for their potential to contribute to those goals. By early 1998, they had completed a draft plan for one of the 28 watersheds, Neponset, which flows into Boston Harbor from the south. Watershed needs and deficits had been defined, more than 160 potential restoration sites (individual, not banking sites) were identified, and finally, the community had been involved in the establishment of seven goals for the watershed and the selection of 40 priority sites (personal communication with Christy Foote-Smith, April 29, 1998).

In the Puget Sound region of the state of Washington, two state agencies are attempting to create watershed-based programs, in separate efforts, to drive their wetland restoration programs: the Department of Ecology and the Department of Transportation. The region contains 18 river systems which, in turn, contain approximately 200 watersheds. The Department of Ecology has recently completed the development of a Wetlands Restoration Plan and the establishment of a database for one of these river basins, the Stillaguamish. The planning document defines river basin problems, identifies wetland functions that address those problems, locates potential wetland sites, characterizes the wetland potential for providing key functions, assesses the restoration potential of the identified sites, and establishes a qualitative rank for each function. The Stillaguamish database provides a detailed description of wetland restoration sites and the functions each has the potential to provide, if restored. The second state effort is the development, by the Department of Transportation's Environmental Program, of a watershed approach to the selection of wetland mitigation alternatives and the employment of wetland banking as an alternative to the traditional project-specific siting. The new approach is being applied to mitigation project development in the Snohomish watershed, another of Puget Sound's 18 river basins. Neither of these programs has been used yet to site actual mitigation projects.

SUCCESS CRITERIA FOR WETLAND RESTORATION

Wetland mitigation and even mitigation banking have become essential ingredients in our national wetland policy today, and it is probably time to put aside the debate over whether or not mitigation can truly compensate for wetland functions and values lost to economic development activities. Economic development is, after all, the way that our country and its people have survived, and there is no indication that it is decreasing in national importance. The attention can be more productively focused, instead, on how compensatory restoration can be made to work best. The first, most important step in that process is to understand what we want it to accomplish; in other words, to define the criteria for success.

Leaders in the field seem to agree, today, that successful restoration projects must replace the important functions of the lost wetlands, which will by necessity be those the loss of which cannot be avoided or minimized in the development activity. Where the understanding of those functions or the technology to replace them are lacking, loss should be avoided or, at the very least, minimized. In determining the degree to which the understanding and the technology are available, the following questions could be asked first, following from J. J. Ewel's 1987 suggestions (Ewel et al., 1987), which are quoted in the Wetlands chapter of the National Research Council's 1992 *Restoration of Aquatic Ecosystems*:

- Can we sustain the ecosystem?
- Can we avoid invasion by exotics?
- Can we generate a productivity similar to a natural counterpart?
- Can we retain nutrients?
- Can we cause biotic interactions similar to those of reference systems?

If we can do these things at all, then we can identify specific success criteria for each ecological objective, that is, spell out the indicators of the ecosystems we want to protect, the exotics we want to repel, the productivity we will expect from our restored environment, the level of nutrient retention we hope to achieve, and the specific biotic interactions we intend to stimulate.

Hydrologic criteria must also be established, criteria for such functions as groundwater recharge, shoreline stabilization, flood-peak reduction, tidal flow restoration, resilience during hydrologic and climatic fluctuations, erosion control, and wave action reduction. Criteria relative to soils, their texture, and their organic content may be important. Some sites may provide opportunities for educational, aesthetic, or recreational goals to be met, and others may require resistance to human disturbances. Each criteria set will be specific to the individual project.

Where compensatory restoration appears to be possible, and realistic goals (success criteria) have been established, it would then be necessary to do the following:

- Assess the structural and functional attributes of the wetlands being destroyed in a temporal context and in the spatial context of the physical land and water regimes and the regional ecosystems. This requires some form of watershed planning.
- Set project-specific success criteria tied to those attributes (see above). Select a mitigation site with the best potential for satisfying those criteria. Establish a schedule for frequent monitoring for accomplishment of those goals and a mechanism for making changes, both during the construction of the restoration project and for a period of time after its completion.
- Do a rigorous project evaluation in terms of the previously identified success criteria no sooner than 5 years after project completion.

What this means, simply, is to do the planning in a broader context than is usually used, to be flexible as the project develops, and to extend the monitoring beyond the usual cutoff points. If intelligent, well-trained people on both sides of the regulatory bargaining table give thoughtful consideration to each of these steps, are flexible enough to make changes as they progress, and have the time and money to follow the project into the future, the nation may quite comfortably meet its wetland policy goals.

REFERENCES

Dahl, T. E. and C. E. Johnson, *Wetlands: Status and Trends in the Coterminous United States, Mid-1970s to Mid-1980s*, First Update of the National Wetlands Status Report, U.S. Department of the Interior, Fish and Wildlife Service, Washington, DC, 1991.

Dennison, M. S., *Wetland Mitigation: Mitigation Banking and Other Strategies for Development and Compliance*, Government Institutes, Rockville, MD, 1997.

Ewel, J. J., W. R. Jordan III, M. Gilpin, and J. Aber, "Restoration is the Ultimate Test of Ecological Theory," *Restoration Ecology: A Synthetic Approach to Ecological Research*, W. R. Jordan III, M. E. Gilpin, and J. D. Aber, editors, Cambridge University Press, Cambridge, 1987.

Foote-Smith, C., personal communication, April 29, 1998.

Frayer, W. E., T. J. Monahan, D. C. Bowden, and F. A. Grayhill, *Status and Trends of Wetlands and Deepwater Habitats in the Coterminous United States, 1950s to 1970s*, Colorado State University, Fort Collins, CO, 1983.

Heimlich, R. E., K. D. Wiebe, R. Claassen, and R. M. House, "Recent Evolution of Environmental Policy: Lessons from Wetlands," *Journal of Soil and Water Conservation*, May–June, 1997.

Kryzinski, W. A., "Wetlands and the Section 404 Program: A Perspective," *Wetland Creation and Restoration, The Status of the Science*, Volume II, USEPA, 1989.

Kusler, J. A., "What Is Wetlands and Watershed Management? Why Is It Needed?" and "Key Issues, Steps, and Procedures In Wetlands and Watershed Management," *Wetlands and Watershed Management: A Collection of Papers from a National Symposium and Several Workshops*, J. A. Kusler, D. E. Willard, and H. C. Hull Jr, editors, Institute for Wetlands Science and Public Policy, The Association of State Wetland Managers, Berne, New York, 1995.

Kusler, J. A. and M. E. Kentula, editors, *Wetland Creation and Restoration: The Status of the Science*, Volumes I and II, USEPA, 1989.

Lewis, R. R. III, *National Wetlands Newsletter*, Volume 19, Number 4, July–August, 1997, p. 22.

Lewis, R. R. III, "Wetlands Restoration/Creation/Enhancement Terminology: Suggestions for Standardization," *Wetland Creation and Restoration, The Status of the Science*, Volume II, edited by J. A. Kusler and M. E. Kentula, USEPA, 1989.

Shabman, L. and P. Scodari, "Wetlands Mitigation Success Through Credit Market Systems," *Wetlands and Watershed Management*, J. A. Kusler, D. E. Willard, and H. C. Hill, Jr., editors, The Association of Wetlands Managers, Berne, New York, 1995.

Tolman, J., "How We Achieved No Net Loss," *National Wetlands Newsletter*, Volume 19, Number 4, Washington DC, July–August, 1997.

United States General Accounting Office, *Ecosystem Management: Additional Actions Needed to Adequately Test a Promising Approach*, Washington, DC, 1994.

United States Environmental Protection Agency, Fact Sheet #26, February, 1995.

Vanderpool, G., "Wetland Banks Gaining Acceptance," *Mississippi Monitor*, Washington, DC, March, 1998.

CHAPTER 6

FLORIDA KEYS BRIDGE REPLACEMENT

One way to relieve the tedium of northeastern United States winters in the mid-1960s was an occasional dash to the Florida Keys. The grueling nonstop (and pre-interstate) drive down the coast, sometimes interrupted by the traffic cops that lingered behind billboards at the entrances and exits of the many small southern towns, was well compensated by the exquisite pleasure of entering the brilliant and fragrant paradise that began just beyond Homestead in southern Florida. From a first stop at the bar where Humphrey Bogart's *Key Largo* was filmed, through the miles of uninhabited mangrove swamps, past the small settlement at Marathon and clusters of rough cottages on Big Pine Key, to the roaring party hosted by after-dark Key West, the drive was across bridge after bridge after bridge—all narrow, many rickety, yet offering glorious vistas of sparkling sea to the left and a mysterious maze of channels and islands to the right. One fished off those bridges, rowed and fished and snorkeled under them, and always marveled at their ability to sustain the weight of the steady stream of traffic. Crossing Seven-mile Bridge, particularly, was an act of courage, part of the cherished experience that was the Florida Keys.

In 1976, the Florida state legislature authorized the Florida Department of Transportation (FDOT) to replace 41 of the 42 bridges that connected the string of islands off the southern tip of Florida known as "the Keys." It was generally agreed at the time that the condition of these 41 bridges made travel from Miami to Key West along U.S. Highway 1 unreasonably slow and that their structural deterioration made it dangerous as well. Those interests that used the highway for business and recreation supported the project. It was opposed by others who feared that the construction, as well as the influx of new visitors that it would produce, would damage or even destroy the unique Keys environment and ecological systems. The two sides of this argument were represented within state government by the FDOT supporting the bridge replacements and the Department of Environmental Regulation (DER, today the Department of Environmental Protection) opposing it. Other natural resource agencies, both state and federal, lined up on the side of the DER, but the FDOT had the advantage of a legislative mandate and an appropriation of the money to do the job.

The project survived the National Environmental Policy Act (NEPA) process with a negative declaration, and the Corps' permitting process (Section 404 of the Federal Water Pollution Control Act, or FWPCA) was not yet fully operational. The FDOT could not proceed with the project, however, without obtaining Florida state water quality and dredge and fill permits from the newly created DER. The DER first took a stance in opposition to *any* new construction on the Keys highway, and early negotiations between the DER and the FDOT consisted of the agency staffs just "duking it out." The atmosphere was combative, exacerbated by pugilists who, for example, began one negotiating session by putting a sign on the table: "Hungry? Eat an environmentalist." The transportation people took the position that they were going to build what they wanted, where they wanted, and the natural resources agencies responded with the bureaucratic equivalent of "over my dead body."

Eventually negotiations focused on specific questions: the alignment of the new bridges; whether new bridges or fill with culverts should replace the old bridges; how to avoid the destruction of mangroves and seagrass communities; habitat losses, water quality degradation, reduction in water exchange between the bodies of water on either side of the Keys; and where to put the spoil. Although the concept of mitigation for environmental damages was relatively new, eventually it became part of the dialogue.

Design changes were negotiated, construction and disposal techniques were agreed upon, and mitigation requirements were endorsed in the agreement that was finally crafted by the FDOT and the DER, on December 13, 1976, requiring turbidity controls during construction, compensatory plantings of mangroves, and a procedure by which seagrass losses might be mitigated. With the signing of the agreement, which was incorporated into the DER permits and, subsequently, into the 404 permits that were eventually required, the way was clear for the construction of the bridges. One by one, beginning in 1978, the permits were issued and construction began. By 1982, 37 bridge replacements had been completed.

Accomplishing the mitigation, however, was more difficult. Lost mangrove and seagrass communities had to be replaced and the approaches to the bridges had to be revegetated. Once the mangrove plantings were underway it was clear that there was insufficient right-of-way to accomplish the required amount of mitigation. FDOT personnel began a search for additional sites under public ownership. The opportunity to test a new concept was discovered on Bahia Honda Key, where tidal connection to an interior lagoon had been severed. The

FDOT made a cut to restore tidal circulation and subsequently, when the revitalization of the lagoon's ecosystems appeared to be underway, they installed culverts at three locations on Boca Chica Key to provide for tidal circulation that was expected to restore and expand a much larger area of mangroves.

Very little was known, at the time, about seagrass restoration techniques. Some experimental research was done on seagrass planting at Craig Key, the results of which, according to one participant, were subsequently ignored. The FDOT eventually wrote a check to the DER in lieu of providing compensatory seagrass plantings. The DER, in turn, contracted with a private consultant to conduct workshops that did some experimental plantings of turtle grass (*Thalassia testudinum*) in the damaged areas. Some past participants in those workshops chuckle today in recollection of the purported "great *Thalassia* spill"—an escape of quantities of turtle grass seed that was blamed for the lack of control plots in a controversial piece of research; others shiver on being reminded of the unseasonable cold in which they donned jeans and sweatshirts (one, with a bullet hole in it, picked up in a flea market) over their wetsuits as they submerged to plant *Thalassia* plugs. An evaluation of the program, conducted in the mid-1980s, indicated that it was relatively unsuccessful. A third effort to mitigate environmental damages was the seeding of those slopes on the bridge approaches where natural revegetation was not occurring.

In 1993, the FDOT funded an evaluation of the project that documented, in a publication the following year and to the surprise of many involved, that the tidal exchange accomplished by the installation of the Boca Chica culverts had so revitalized the interior lagoons that not only had substantial enhancement of mangrove communities occurred, but new seagrass meadows had appeared as well (Lewis et al., 1994). Although earlier documentation made it difficult to provide complete comparisons, observations and photographs taken in 1993 recorded the successful development of slope revegetation and shoreline mangrove plantings at many of the bridge locations. The 1993 survey was able to report, therefore, the satisfactory accomplishment of compensatory restoration of both the seagrass and the mangroves lost because of the bridge replacement construction.

ENVIRONMENTAL AND SOCIAL SETTINGS

The Florida Keys are exceptional in the complexity and diversity of the natural environments they contain and in the unique coral bank reef that parallels them on the ocean side. The high productivity of these marine environments and the fragility of the coral reefs deserve vigilant protection from the impacts of development, yet it is these very special natural resources, along with the recreational opportunities, temperate climate, and physical beauty of the area, that stimulate the development associated with population growth and tourism.

The Keys are a limestone island archipelago extending approximately 190 miles southwest from the southern tip of Florida to the Dry Tortugas (see Fig.

Figure 6-1 Project location.

6-1). They are made up of more than 1,700 virtually flat islands, a small portion of which is inhabited. The easternmost inhabited island is Soldier Key, the westernmost is Key West. The Keys are divided into the Upper Keys, from Soldier to Lower Matecumbe Key; the Middle Keys, from Lower Matecumbe to the Seven Mile Bridge; and the Lower Keys, from Little Duck Key to Key West. A fourth morphological area comprises the Marquesas and Dry Tortugas, those uninhabited and isolated islands that lie well beyond Key West. The Upper and Middle Keys are elongated, paralleling the axis of the chain, and are composed of Key Largo limestone. This 120,000-year-old former coral reef extends below Miami, Florida Bay, and the Dry Tortugas and surfaces from Soldier Key to the Newfound Harbor Channel, ranging from 23 to 52 m thick. It has high porosity and permeability.

The Lower Keys, on the other hand, are broad and flat, separated by narrow channels with their axes perpendicular to the axis of the chain. They are composed of Miami oolite, a series of fossilized sandbars that developed concurrently with the Key Largo limestone and, beginning at Big Pine Key, overlie it. Though also highly porous, the oolite is much less permeable than the limestone. Surface sediments are mainly marine calcareous sands. Peat is sometimes found in depressions, and limy marls occur infrequently (Kuyper, 1979).

The Keys are located at the southern edge of the carbonate platform known as the Florida Plateau, onto which sedimentation has been occurring for 150 million years. The 7,000-m thick plateau is underlain by the crystalline and sedimentary basement rocks of the South Florida Basin, a block-faulted feature associated with the breakup of North America and Africa during the Mesozoic era. The region's current geomorphology is attributed mainly to sea level fluctuations caused by Pleistocene glaciations, terminating with the increase in sea level that permanently covered the area during the Wisconsin glaciation about 6,000 years ago.

The triangle of water between southern Florida and the Lower Keys is called Florida Bay (Figures 6-2a and 6-2b), a shallow area filled with mud flats composed mostly of calcium carbonate. The bay experiences wide fluctuations in temperature and salinity and periods of high turbidity. On the opposite, southeast side of the archipelago is the Atlantic Ocean and the Florida Reef Tract, a series of living coral bank reefs that parallel the keys themselves. The Florida Reef Tract is located on a narrow shelf that drops off, further out, into the Straits of Florida. This tract comprises one of the largest communities of its type in the world; bank reefs extend for 130 km from the Marquesas to near Miami and are edged with approximately 6,000 patch reefs. The warm nutrient-deficient waters resulting from the tidal exchange with the Atlantic Ocean are important to healthy reef development.

> Both patch and outer reefs maintain a balance between physically constructive elements (including corals, algae, and other flora) and destructive elements (e.g. salinity and water temperature changes, turbidity due to weather events, exposure to air, and changes in nutrient levels). By altering the physical characteristics of the reef environment, human activities may further stress an already stressed ecosystem.
>
> —NOAA, 1995 (through Jaap and Hallock, 1990); Voss, 1988

The reef tract is separated from the islands of the Keys by Hawk Channel (see Figs. 6-2a and 6-2b), which itself contains a coral reef and its biotic communities. Movement of nutrients, sediments, salinity, and temperature through the channels between the Keys occurs primarily in the direction from the bay to the ocean side. Reef development is greater off the Upper and Lower Keys than off the Middle Keys.

Circulation of waters along the Atlantic side of the Keys is dominated by the Loop Current which, entering the Gulf of Mexico from the southwest, loops north and clockwise and exits southwest of the Dry Tortugas to become the Florida Current, traveling along the state's east coast in a northerly direction. This water movement typically produces *gyres* along its path, counterclockwise currents of cold water that have been observed to trap nutrients along the bank

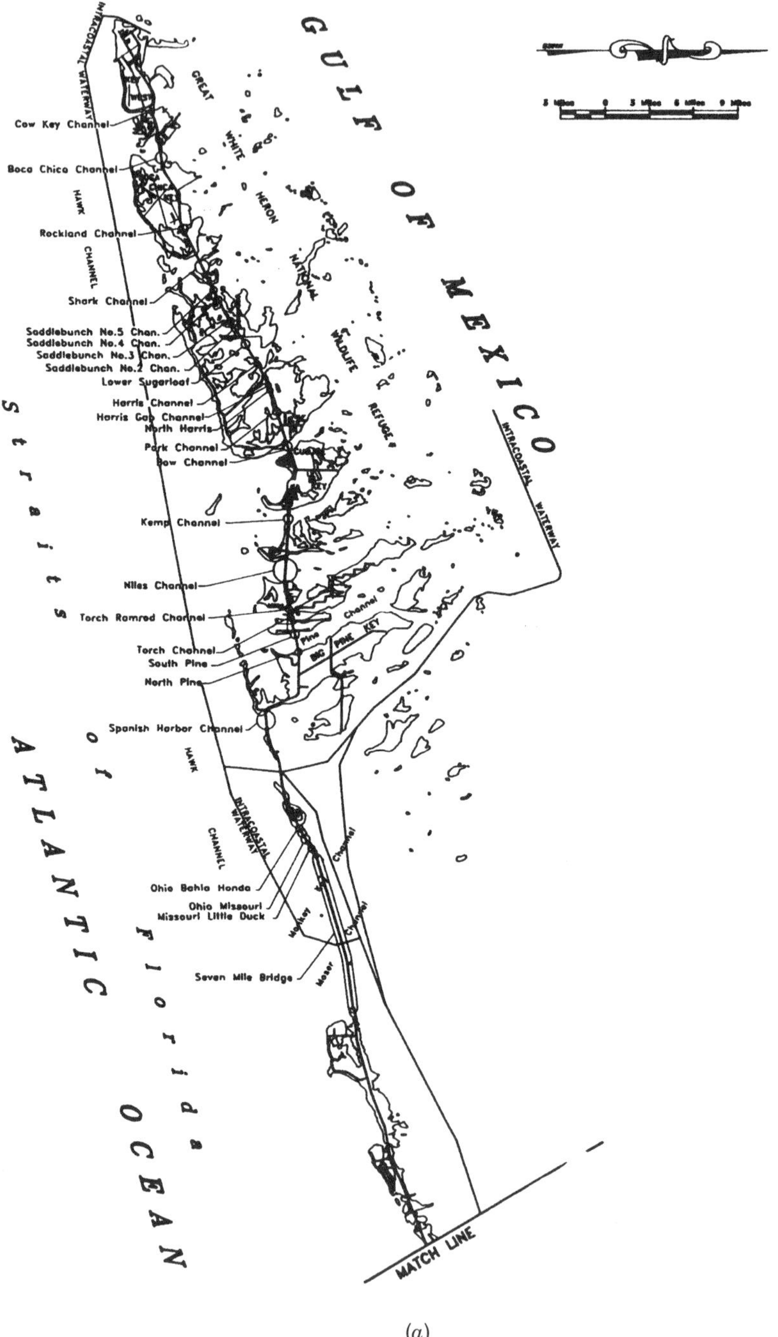

(*a*)

Figure 6-2a Lower Keys bridge locations. Reproduced with permission of Lewis Environmental Services, Inc. from *Wetlands Mitigation Evaluation Report Florida Keys Bridge Replacement*, 1994.

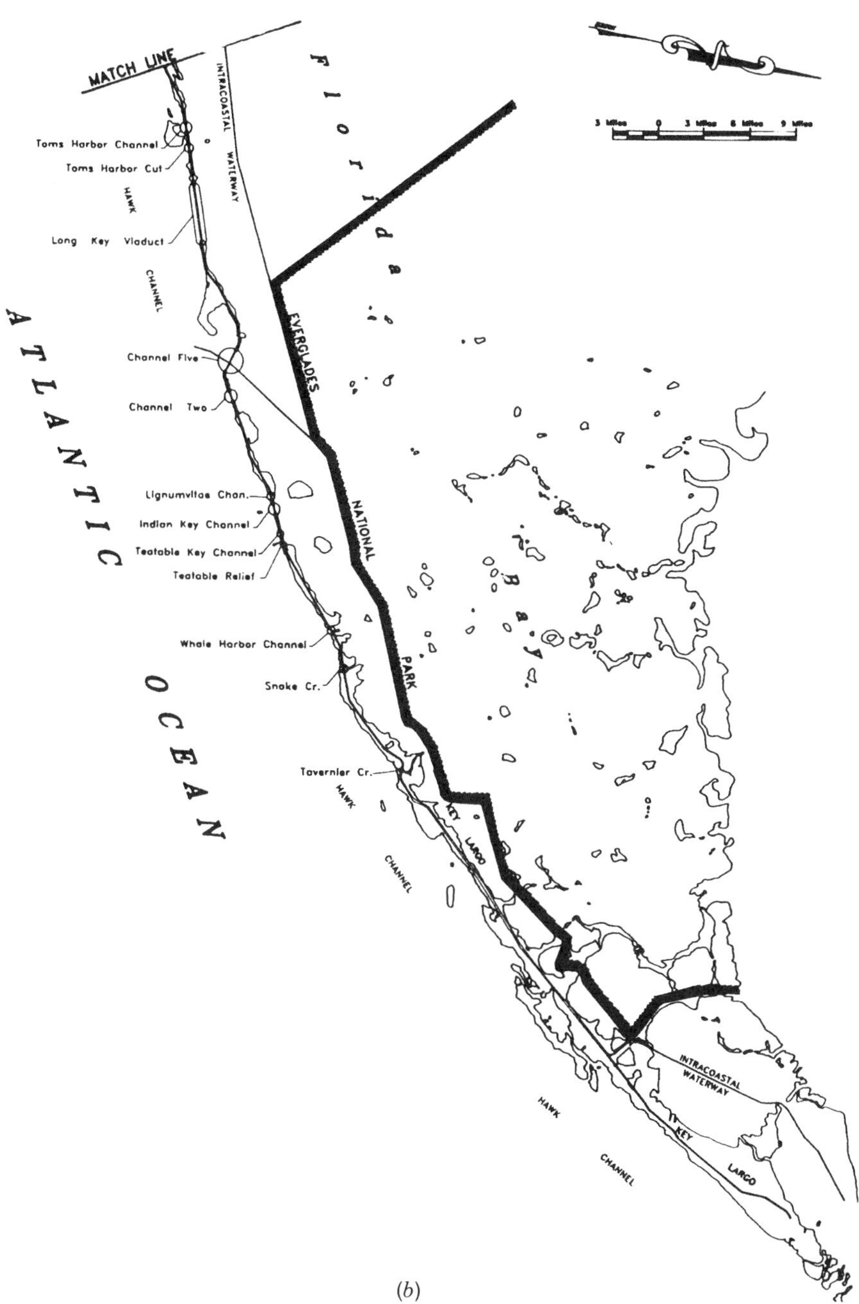

(b)

Figure 6-2b Upper Keys bridge locations. Reproduced with permission of Lewis Environmental Services, Inc. from *Wetlands Mitigation Evaluation Report Florida Keys Bridge Replacement*, 1994.

reefs near the Lower Keys and provide flushing to the reefs of the Upper Keys (NOAA, 1995). The warm waters of the Florida Current are essential to the survival of the Florida Reef Tract.

The little rain that falls in the Keys is quickly absorbed into the porous rock formations and ends up in shallow lenses, most of which are brackish. There are no freshwater springs on the Keys although a few of the larger islands have fresh water in lagoons and subsurface lenses. Groundwater in South Florida and the Keys consists of both these lenses and the 150- to 300-m-deep Floridan Aquifer, which underlies not only Florida but also portions of Georgia, South Carolina, and Alabama. The groundwater in this aquifer system ranges from brackish to seawater and it is used mainly for subsurface storage of liquid wastes.

The region has a tropical maritime climate with moderate temperatures and two seasons: a long, wet summer extending from May to October, and a mild, dry winter from November to April. The circulation patterns of the Florida Current and Gulf Stream influence the climate, as do also the warm waters of the Gulf and the Atlantic Ocean. The weather is associated with the tropical maritime air of the Bermuda/Azores high pressure system, and tropical storms and hurricanes are typical features. The annual probability of a hurricane occurring is between 13 and 16%, and Key West has averaged one hurricane every eight years. The 1935 Labor Day storm in the Keys is the most violent to have made landfall in the United States and is one of only two Class 5 hurricanes ever to have hit the United States. Winds are from the east–southeast during the summer and the east–northeast during the winter. The flatness of the terrain makes the Keys particularly vulnerable to the damage of hurricanes; 96% of the land mass is less than 2 m above sea level and the highest elevation, on Windkey Key, is only 18 ft. Storm surges, produced as underlying water domes up in response to low air pressure and combines with high wave action, are considered the greatest threat to the Keys. The Lower Keys have the nation's greatest frequency of occurrence of waterspouts.

Precipitation across the Keys is lower than in the rest of Florida, greatest in September (16.5 cm monthly mean) and least in March (3.3 cm), with an average annual rainfall of 124.5 cm. Most rainfall occurs in the form of the intense summer storms. Temperatures are the most moderate in the state, varying little either across the Keys or throughout the year. Temperatures in Tavernier and Key West are typically within 1°C of each other. Average annual temperatures at Key West vary from a maximum of 28°C to a minimum of 23°C. The mean average annual relative humidity is 75% and remains constant throughout the year. Droughts are most likely to occur during May, June, September, and October.

The Keys' ecosystem is considered to be ecologically and aesthetically unique within the United States and is part of a wider system described as "one of the most complex ecosystems on Earth" (NOAA, 1995). The larger ecosystem ties the nearshore habitats and tidal channels of the islands themselves to the ecosystems of the lower Everglades, Florida Bay, the Gulf of

Mexico, and the Atlantic Ocean. The nearshore habitats, including the important mangrove and seagrass environments, provide shelter, food, and nurseries for many of the species of the broader area, and the human activities on the Keys, including such infrastructure projects as the Key Bridges Replacement, have substantial impacts on water quality, specifically upon the clarity of water so critical to the Florida Reef Tract offshore.

A wide range of habitats are available on the Keys themselves, including beaches, coral reefs, seagrass meadows, pine rockland, transitional wetlands, freshwater wetlands, mangrove and hardwood hammocks. Two distinct marine habitats can be observed on either side of the archipelago: the Gulf of Mexico's warm-temperate habitat in the Florida Bay and a tropical Caribbean habitat on the Atlantic side, with an ecological and biological mixing zone in the nearshore area of the islands themselves. The Florida Reef Tract comprises one of the largest communities of its type in the world. In addition to the bank reef habitat, it includes: offshore patch reefs, seagrass, back reefs, bank reefs, intermediate reefs, deep reefs, outlier reefs, and sand and soft bottom.

Seventy-one species of plants in the Keys are listed as threatened or endangered by the Florida Department of Agriculture and Consumer Services (FDA). The U.S. Fish and Wildlife Service (FWS) has identified two species as federally endangered, Key tree cactus (*Cereus robinii*) and Small's milkpea (*Galactia smallii*), and one as federally threatened, Garber's spurge (*Euphorbia garberi*).

Among the animals that have been classified as threatened or endangered by either the federal or the Florida governments are the following:

- Invertebrates: Florida tree snail, the Stock Island tree snail, the pillar coral, Schaus's swallowtail butterfly.
- Fishes: the common snook, Key blenny, Key silverside, and mangrove rivulus.
- Amphibians and reptiles: the American alligator and the American crocodile; the Atlantic green, Atlantic hawksbill, Atlantic loggerhead, Atlantic Ridley, leatherback, and striped mud turtles; the Big Pine Key ringneck, eastern indigo, Florida brown, Florida ribbon, Miami back-headed, and red rat snakes; and the Florida Keys mole skink.
- Birds: the American kestrel, American oystercatcher, Arctic peregrine falcon, Bachman's warbler, brown pelican, burrowing owl, Cape Sable seaside sparrow, Florida sandhill crane, least tern, little blue heron, snowy egret, tricolored heron, osprey, piping plover, reddish egret, roseate spoonbill, roseate tern, Southeastern snowy plover, white-crowned pigeon, and wood stork.
- Mammals: blue, fin, humpback, right, sei, and sperm whales; Florida manatee; Key deer; Key Largo cotton mouse; Key Largo wood rat; silver rice rat; and Lower Keys marsh rabbit.

The Keys today are particularly attractive for retirement and seasonal living and as a vacation destination, and the permanent population is on the rise. Human interface with the Keys environment, however, creates special problems on both sides: the absence of fresh drinking water, the poor facilities for disposing of waste water, the damaging effects of recreational activities on the more fragile elements in the environment, and the damages caused by the infrastructure necessary to support human populations are a few examples, but they are all fairly recent phenomena.

Little is known about prehistory in the region. Although there have been no specific finds (most of the likely sites being underwater sinkholes), it is believed that "the Keys have an excellent prospect for human and animal remains that are between 12,000 and 15,000 years old" (NOAA, 1995). Earliest European interest in the Keys was limited to concern for the hazards they presented to navigation. A lightship was first stationed in the Keys in 1824, followed by the 1825 start of construction of a permanent lighthouse in the Dry Tortugas. More recently, most of the historical interest in the Keys has centered around shipwrecks.

The Florida Keys were first connected to the mainland by Henry Flagler's Overseas Railroad, constructed between 1904 and 1912. This construction included the erection of 27 km of bridges across open water and 32 km of filled causeways. The following quote from *Florida: The Long Frontier* gives some of the flavor of the event and a sense of the disregard for environmental values at that time:

> Steadily and slowly from 1904 to 1912 Mr. Flagler's great project of the Overseas Railroad crept south and west in blazing summers and damp winters, surviving three hurricanes. Swarms of vessels, work boats, dredges, seagoing cranes, pile drivers, quarter boats, floating cement mixers, left the Miami River docks to join the work. . . . More people came in to work on the Key West Extension, as wooden bridges and solid causeways crept from Key to Key, on sunken piles and rock fills of cement that would harden underwater, on arches bristling with reinforcing rods striding out across the quick-flowing channels. Embankments were made of 240 acres of rock and sand reclaimed from the sea. Work camps were set up on the cleared wastes of key beaches. The blazing days rang with the shouts and clangor, the dynamite explosions of progress. Nights glared with lights on poles, electricity supplied by dynamos on scows, thumping like the heartbeats of progress. In all, more than three or four thousand men, Negro and white, swarmed over the advancing roadways, viaducts, bridges. As the water reaches widened to the south, men lived in floating quarter boats, fed and watered by swarming supply boats and boats of busy peddlers. . . . The hurricanes that came howling across the gray waves to the advancing line of arches in 1906, 1909, and 1910 washed away roadbeds, drowned men, and hampered the work. . . . In January, 1912, Mr. Flagler, eighty-two years old, arrived on the first train, all the way from Jacksonville, rumbling over bridges and viaducts and the sea, into the once-isolated island of Key West. The city was beside itself with excitement. . . . The Overseas Railroad completed his great work of opening up the east coast of Florida.
>
> —Douglas, 1967

When a portion of the railroad was destroyed by the 1935 hurricane, the state of Florida purchased the 156-mile stretch between Homestead and Key West and began construction of the Overseas Highway on the railroad right-of-way, a roadway that is the conduit of entry to the islands today, along with two public airports, four private airstrips, and inestimable amounts of water traffic.

The Keys lie within Monroe County and include three incorporated areas: Key West, Layton, and Key Colony Beach. Although the inhabited Keys make up only 5% of the land area of Monroe County (65,500 of 1.2 million acres, including 99,000 acres in the Everglades), most of the county's population is located in the Keys. The 1990 population was estimated to be 134,600; 78,000 permanent residents and the remaining seasonal. About one-third of the population is located in the three incorporated areas. An additional 2,500 "live-aboards" are estimated to reside on the boats anchored in the more than 160 marinas in the Keys. Key West was home to 90% of the year-round population of Monroe County prior to the 1940s, but to only 32% by 1990. Approximately 30% of the population is over 55 years old, higher than the national average.

By the mid-1970s the crush of new development projects was seen as a threat to the natural resources of the area, and in 1975, Florida designated Monroe County as an Area of Critical State Concern, thus assuming state authority over planning and development activities. One of the outgrowths of state oversight was the establishment of the Monroe County Land Authority to acquire designated large natural areas. Just over 34% of the Keys' total land area is vacant, making 210,000 acres available for development. Current land use patterns vary according to the location. Residential land use ranges from 12% in the Lower Keys, where there is a high percentage of conservation land, to 58% in Key Colony Beach. Public facilities and buildings account for 8% of Key West and 1% or less of other areas. Military Buildings make up 25% of Key West's land use, and in the remainder of the Lower Keys there are three military facilities that make up 5% of all land in the unincorporated area. About 4% of land acreage within the Keys is in commercial use. Rock mining and marine repair and salvage are the only Keys industries.

Although the economy is dominated by tourism, the military and commercial fishing are also important. In 1990, tourism generated 16,691 jobs, half of all employment in the county. Service and retail industries are the largest private sector employers; finance, insurance, and real estate trades are the second most important employment sector. Commercial fishing is the fourth largest employment sector, accounting for 9% of the work force. Public sector employment is 23% of the total work force: 7% military; 9% state and county, and 7% federal. The military is a particularly important employer in the Key West area.

Recreational facilities make up 7% of Key West land use; less than 2% in the Lower and Upper Keys, and 11% in the Middle Keys. There are a total of 257 public and private recreation sites, including 24 federal and state facilities, among them being the John Pennekamp Coral Reef State Park, Bahia Honda State Park, Looe Key National Marine Sanctuary, Key Largo National Marine

Sanctuary, the National Key Deer Refuge, the Dry Tortugas National Park, Key West National Wildlife Refuge, Great White Heron National Wildlife Refuge, Fort Zachary Taylor State Historic Site, and Long Key State Recreation Area. Conservation land constitutes 34% of all unincorporated land, the largest refuges being the National Key Deer and Great White Heron refuges. A broad area that includes the Keys was designated by Congress as a National Marine Sanctuary by the 1990 Florida Keys National Marine Sanctuary and Protection Act. The Act gives certain planning and management responsibilities to the National Oceanic and Atmospheric Administration (NOAA) in the U.S. Department of Commerce. The process initiated by this designation identified four critical management issues that needed to be addressed: the decline of water quality, physical injury to resources, decline of marine resources, and use conflicts.

Human impacts upon the water quality of the Keys include point and nonpoint sources. Point sources include 10 domestic wastewater treatment plants, the largest of which is the Key West Sewage Treatment Plant. The most important nonpoint source pollution comes from domestic wastewater, and the largest source of domestic wastewater is the on-site disposal systems (OSDS). In 1992, the USEPA estimated the existence of 30,000 septic tanks and cesspits in the Keys.

THE PROJECT AND ITS WETLAND IMPACTS

Land access to the Keys is restricted to the single highway, U.S. 1, the Overseas Highway, which links the islands together by a series of 42 bridges between Homestead, on the mainland, and Key West, the furthermost populated island. Even today, only 51 of the 1,700 islands in the Keys are connected to or by U.S. 1. By the 1970s the bridges had deteriorated sufficiently that repairs and maintenance were costing the FDOT as much as their replacement cost, and in the early 1970s they began to develop a plan to accomplish that replacement. The two reasons put forward by the FDOT were safety—people were actually afraid to drive on some of the bridges, and the drawbridge span on Seven-mile Bridge got stuck frequently—and ease of evacuation in case of a hurricane. The FDOT did not perceive it as a development issue, because the mostly two-lane bridges were not being enlarged. The Florida legislature authorized the Keys Bridge Replacement Program at an estimated cost of $175 million in 1976. Because so many bridges were involved, the concerned state and federal agencies agreed early on that negotiating the conditions of the necessary environmental permits would be done as a single program, although individual permits were eventually issued for each of 37 bridges to be replaced. Ultimately, the environmental permits were conditioned on the relatively brief requirements contained in a Memorandum of Agreement (MOA) signed by the FDOT and the DER on December 13, 1976.

The FWPCA authorizing the Corps' Section 404 dredge and fill permitting process had been passed only 4 years prior to the legislative authorization for the bridges and, although Section 404 permits were required by the time actual bridge construction began, they were not an important consideration when the negotiations began. The primary permit that the FDOT had to have at that time was from the Florida DER, which itself had just been created in 1975 by combining the former Department of Pollution Control with other related state functions. The new state environmental agency had no established regulations or procedures and it was agreed that a single program would be developed to address the potential environmental damage from all the bridge replacements by negotiation among the agency staffs. The major players for the State of Florida were the FDOT, the DER, and the Department of Natural Resources (DNR). The involved federal agencies were the Corps of Engineers, the FWS, the National Marine Fisheries Service (NMFS) (both being, at that time, in the U.S. Department of the Interior), and the U.S. Coast Guard.

The original intention of the transportation engineers was to replace as many of the bridges as possible with filled-in causeways, installed with culverts for water circulation, a technique that would have been cheaper and made repair and replacement easier, after storm damage. They argued highway safety, increased employment for the region, and storm evacuation benefits—there was even talk of a military evacuation involving the landing of C-130s on Highway 1—and they had the advantage of a legislative authorization and millions of appropriated dollars on their side. The DER objected strongly to causeways, being concerned about the many restrictions to flow circulation and biotic transgress between the Atlantic Ocean and the Florida Bay. They were concerned about the project altogether because of the impacts on water quality from the construction and the increased slope erosion. Their preference, if any work at all had to be done, was for bridges only, to be as closely aligned as possible with the original ones. The position of the resource agencies was, according to one engineer who participated in the process, "just don't build there . . . total avoidance of the issue."

There were various alignment issues related to individual bridges. Alignment was complicated by the presence of the aqueduct that provides the Keys with their supply of fresh water; it was constructed beneath the old bridges and would have to be transferred to any new ones. The DNR and the FWS were concerned about loss of habitat as a result of the destruction of submerged aquatic vegetation—seagrass—and the mangroves that lined the shores of the islands. They argued for project design and construction procedures that would minimize these impacts and for compensatory plantings that would replace the lost vegetation. All the natural resources agencies were concerned about increased turbidity, both during the construction itself and as a consequence of the constructed nonvegetated and unstabilized slopes. The question of the impacts of construction turbidity on seagrass communities was argued in the scientific community and in the newspapers. The FDOT commissioned a 3-year research project at Seven-mile Bridge that concluded that thunderstorms in the keys did "a lot more" damage to turtle grass beds than did dredging operations (*Key West Citizen*, May 27, 1982). A related issue was disposal sites for the considerable debris that would be generated by the destruction of the old bridges. There was no discussion of mitigation ratios and no formal mitigation policy to fall back on.

The agencies argued about construction techniques and finally agreed that the contractors would use the box girder segmental design system, building the bridge away from itself, which was new in the United States at that time. They cast the segments elsewhere and brought them in by barges pushed by tugs. (The DOT applied for permits for two design schemes: the segmental and the traditional one, and then used the appropriate permit after they decided on the design.) Alignment issues were resolved, to neither side's complete satisfaction. The NMFS unsuccessfully asked that backfilling and seagrass planting in the work channels be part of the permit conditions. In the process of negotiating the disposal issue, the nonprofit Keys Artificial Reef Association (KARA) was born, an organization that continues today and that successfully removed 35,000 tons of rubble to six permitted artificial reef sites between 1981 and 1987. The requirements for mitigation of seagrass and mangrove losses were addressed in the December 13, 1976, MOA along with construction turbidity-control conditions. Loss of vegetation on the slopes of the approaches to the bridges was a third mitigation issue that arose and was addressed by one of the Particular Conditions of the DER permits: "All unpaved areas of the bridge approaches . . . shall be stabilized by vegetation or other methods approved by this Department." The Corps permits included, as special conditions, a reference to the December 13th MOA and the statement that "seagrasses shall be replaced in like quantities in accordance with planting methods obtained from the FDOT experimental seagrass planting program." A procedure for DER monitoring of the construction process was established. Beginning with the bridge over Cow Key Channel, construction of the entire 37 bridges was completed between March 1978 and October 1982 (see Table 6-1).

The December 13th MOA claimed to satisfy the DER permit requirements for mitigation pertaining "to the loss and re-establishment of desirable wetland and submerged vegetation adversely affected by the construction process." It established three conditions:

1. Turbidity barriers were to be placed wherever suspended sediments could affect aquatic grasses, and riprap barriers were to be constructed to prevent erosion of fill material beyond the construction area.
2. Red, white, and black mangroves were to be left standing wherever possible; natural reestablishment at the toe of the fill slope was to be encouraged and, in locations of new fill, mangrove revegetation would be accomplished by putting mats of seagrass litter within the swales created by rubble berms constructed just beyond the toe of the slopes. If, after 1 year, this operation had not been successful, then seedlings were to be

TABLE 6-1 Bridges Replaced in the Keys Bridge Replacement Program

Bridge	Length (ft)	Completion Date
Tavernier Creek	320	12/78
Snake Creek	230	7/81
Whale Harbor	720	12/78
Tea Table Relief	270	6/80
Tea Table Channel	700	6/80
Indian Key	2,460	7/81
Lignum Vitae	860	7/81
Channel No. 2	1,760	11/80
Channel No. 5	4,580	10/82
Long Key	12,040	7/81
Tom's Harbor Cut	1,270	5/80
Tom's Harbor	1,460	5/80
Seven Mile	35,830	10/82
Missouri Little Duck	840	6/81
Ohio Missouri	1,440	6/81
Ohio Bahia Honda	1,050	6/81
Spanish Harbor	3,380	80-81
North Pine	660	80-81
South Pine	850	80-81
Torch Channel	880	80-81
Torch Ramrod	720	80-81
Niles Channel	4,490	80-81
Kemp Channel	1,030	80-81
Bow Channel	1,340	80-81
Park Channel	880	80-81
North Harris Channel	430	80-81
Harris Gap Channel	140	80-81
Harris Channel	430	80-81
Lower Sugarloaf Channel	1,260	4/80
Saddlebunch No. 2	660	4/80
Saddlebunch No. 3	760	4/80
Saddlebunch No. 4	900	6/80
Saddlebunch No. 5	900	6/80
Shark Channel	2,090	1/80
Rockland Channel	1,280	7/79
Boca Chica Channel	2,730	11/80
Cow Key Channel	360	3/78

Source: Used with permission of Lewis Environmental Services, Inc. from Wetlands Mitigation Evaluation Report Florida Keys Bridge Replacement, 1994.

planted along the intertidal slope at one seedling to every five lineal feet of shoreline.

3. The losses of submerged marine vegetation (seagrass) would be inventoried by the DOT, who would then attempt to revegetate it "to the extent possible under the advice and supervision of the DER and the Department of Natural Resources" by conducting some research which, if eventually agreed to have been successful, would be duplicated elsewhere.

MITIGATION IMPLEMENTATION AND RESULTS

Seagrasses

The fact was that, although the natural resource agencies were asking the FDOT to replace the lost seagrass acreage, no one knew how to do it. Seagrasses are underwater plants that grow in approximately 500,000 acres of Florida's offshore waters and are abundant in the waters of the southern part of the state. Seven of the world's 52 species of marine seagrasses are found in Florida waters. Of those important in the Keys, the turtle grass (*Thalassia testudinum*) is the most dominant form, whereas shoal grass (*Halodule wrightii*) and manatee grass (*Syringodium filiformis*) appear in much smaller quantities in mixed beds or where conditions prevent dense turtle grass growth.

Seagrass performs important habitat and water quality functions. It is the nursery area for much of Florida's recreational and commercial fisheries and provides food and nursery grounds for a wide variety of Florida's fishes, crustaceans, and shellfish. Five communities of organisms inhabit and depend upon seagrass meadows: epiphytic, epibenthic, infaunal, planktonic, and nektonic organisms. Their water quality functions include the trapping of fine sediments and particles in their leaves and bottom stabilization with their roots. Seagrass produces oxygen and its survival is a function of the available light, sediment depth, and turbulence/exposure in shallow water. The correct depth for successful seagrass plantings is mainly a function of the clarity of the local water, and they require at least 1 m of sediment. Declines in seagrass meadows are expected to produce concurrent declines in dependent marine species.

The actual mitigation for seagrass losses proceeded as follows:

- The FDOT funded a study at Craig Key between 1979 and 1981 to determine the technical feasibility of seagrass restoration to mitigate the losses from the bridge replacements (Continental Shelf Associates, 1982). Plugs, sprigs, and seedlings of turtle grass, shoal grass, and manatee grass were planted in 20 experimental plots; only eight of these retained any of the transplanted matter after 2 years. Among the conclusions that were drawn from the Craig Key research was that replanting seagrasses was technically feasible although under certain conditions could be very expensive—up to $42,500 per acre when scuba divers were used to do the plantings. Of the types of plantings, plugs had the best survival rate but their removal had a negative effect on the donor plots. Sponsors of the study recommended that replacement plugs be limited to shoal grass, because there were more likely to be appropriate donor plots available. The survey described in the 1994 evaluation report observed expanded communities of seagrass at the test sites, but could not relate them to the original plantings and surmised that they may have been the result of natural revegetation.

- A supplemental MOA was signed by the DOT and the DER on October 18, 1982, which stated the following:
  - The DOT had "undertaken a mitigation plan with the expressed purpose of replacing or restoring submerged marine vegetation in like amounts to that adversely impacted or destroyed by the bridge replacement construction activity."
  - Fifty-one acres of seagrasses had been disrupted or eliminated by the project.
  - The DOT had conducted an experimental seagrass planting program, February of 1979 to February of 1981, which included "the compilation of technical and economic feasibility data for further mitigation activities and methodologies."
  - By an April 8, 1977 MOA between the DOT and the DNR, the DNR had agreed to provide, at the DOT's expense, technical assistance on seagrass mitigation sites and methods.
  - Modification of the original program was to be as follows:
    - The DOT would pay $200,000 into the DER Pollution Recovery Fund.
    - The DER would use the money to restore or replace seagrasses affected by the Project or, where it is infeasible, to do other mitigation that would "enhance or benefit the marine and aquatic environment of the Florida Keys."
    - This would constitute satisfaction of the DOT obligations.
- The DOT subsequently wrote a check to the DER for $200,000 in return for which the DER took responsibility for the seagrass mitigation and funded Operation Seagrass for $150,000.
- The Florida Keys Seagrass Restoration Project, "Operation Seagrass" was held in two 1-week sessions, in April 16–23 and August 13–20, 1983. Conducted by Mangrove Systems, Inc., under contract with the DER, each session was attended by approximately 25 people. Participants were trained and then actually did submerged plantings (using both snorkels and scuba equipment) of seagrass in 20 separate Keys locations, covering a total of 33.34 acres. Subsequent plantings by Mangrove Systems included 14.0 acres in Sexton Cove and 0.2 acres in Boog Powell Marina, making a total of 47.54 planted acres of seagrass. In 1984, a 61.3% survival rate of the plantings was recorded.

TABLE 6-2 Seagrass Losses

Cause of Losses	Acres Lost
Project fill	24.59
Project dredging	16.25
Shading caused by project	5.71
Propeller wash and cutting during construction	46.78
Total Losses	93.33

The real extent of seagrass meadows destroyed by the Keys Bridge Replacement Project has been calculated, in the 1994 evaluation report, to have been 93.33 acres (Lewis et al., 1994). The 1985 estimate of only 65.8 lost acres was expanded by the subsequent observation of substantial (27.53 acres) additional damage done by propellers during construction. A total of 24.59 acres was permanently destroyed by fill associated with the bridge construction and an additional 5.71 acres were permanently eliminated by being shaded after project completion. The remaining 62 acres were torn up either by dredging (16.25 acres) or by propeller cuts during construction (45.71 acres, including the 27.53 acres discussed above). See Table 6-2.

The seagrass plantings were not successful, but the losses were mitigated (Table 6-3). A total of 93.33 acres was destroyed, leaving 79.37 acres of that area available on site for replanting. There was an attempt to replace 33.33 acres of this during "Operation Seagrass," but in 1984 it appeared that only 20.43 acres had survived. Ten years later, however, that acreage had expanded to 56.64 acres, presumably by natural revegetation. Attempts to mitigate the losses off site had resulted in another 76.55 acres of observable new seagrasses in 1993, most of which (62.2 acres) was in the Boca Chica lagoons as a result of the tidal connection accomplished by the culvert installations. In retrospect, most of the participants in the early mitigation attempts agreed that establishing

TABLE 6-3 Seagrass Mitigation Efforts

Mitigation Effort	Acres
On site	
Acres of lost seagrass for on-site replanting	79.37
Actually replanted on site	33.34
Survived on site, 8/1984	20.43
Present on site, 9/1993	56.64
Off site	
Survived off site, 9/1993	76.55
Total survived, 9/1993	133.19

the conditions for seagrass to regenerate was more important than doing the actual plantings—where seagrass could grow, it did; where it couldn't, it didn't.

Mangroves

Mangrove ecosystems are among the most productive ecosystems in the Keys, with a variety of important values that stand in contrast to the perception of them as impenetrable, mosquito-infested, and unattractive (which, largely, they are). Their most important functions in the Keys are the following:

- Shoreline stabilization: the trapping, holding, and stabilizing of intertidal sediments, protecting landward habitats from hurricane damage, and mitigating the effects of storm waves.
- Importance to endangered species: mangroves are important habitats to at least seven endangered species (American crocodile, hawksbill sea turtle, Atlantic Ridley sea turtle, Florida manatee, bald eagle, American peregrine falcon, and brown pelican), five endangered subspecies (Key deer, Florida panther, Barbados yellow warbler, Atlantic salt marsh snake, and eastern indigo snake), and three threatened species (American alligator, green sea turtle, and loggerhead sea turtle) in south Florida.
- Importance to sport and commercial fisheries: most of the commercial varieties of fish in the area utilize mangrove habitat in their life cycles, including oysters, blue crabs, spiny lobsters, pink shrimp, snook, mullet, spotted sea trout, gray and other snapper, tarpon, sheepshead, and ladyfish.

Mangroves are not a taxonomic category but, instead, are an ecological group that exhibit certain common characteristics. The features that mangroves share are their development of aboveground aerial roots: stilt or prop roots in the case of red mangroves, air roots or pneumatophores in the case of black mangroves, and lenticels in the lower trunk in white mangroves. These root systems spread horizontally over wide areas but are anchored with few underground roots and no tap root. These aerial roots collect masses of leaf detritus, attract algal communities and large populations of marine fungi, provide protection for a wide variety of invertebrates in the maze of prop roots and muddy substrates under them, and are the habitat for spiny lobster juveniles. There are direct grazing insects and the mangrove tree crab feeding on the leaves, prop roots, and mud algae; filter feeders that live on the prop roots and filter phytoplankton and detritus from the water; mobile invertebrates skimming detritus algae and small animals from the mud and forest flood surface; and carnivores that feed upon all the others: the blue crab (*Callinectes sapidus*), caridean shrimp, snapping shrimp, and penaeid shrimp take shelter and eat there. "From the economic point of view, the pink shrimp (*Penaeus duorarum*) is the most important species associated with mangrove areas (Odum et al., 1982).

The typical sequence of species in the Keys mangrove communities, moving from the water's edge upland, is red mangrove (*Rhizophora mangle*), black mangrove (*Avicennia germinans*), white mangrove (*Laguncularia racemosa*) and, finally, buttonwood (*Conocarpus erecta*) and upland species. All these species are present in the scrub forests that fringe the coasts of the Florida Keys. Although they differ in a number of features, they all exhibit the maze of aerial roots that trap the detritus provided by decayed leaf litter, the basic energy source for food chains, and that provide the protected habitat for juvenile fishes and invertebrates.

Among the factors that control and limit the distribution of mangroves are climate, salinity, tidal fluctuation, and substrate (Odum et al., 1982). They need temperatures above 65°F and they grow best in fine-grained muds in environments with low wave action. Although they do not require salinity, anaerobic sediments, or tidal fluctuation to survive, their ability to withstand these conditions gives them a strong competitive advantage in environments where such are present.

The inventory of mangrove losses that the FDOT was required to do showed 51.03 acres permanently lost to fill or excavation. This had to be mitigated, according to the general terms of the December 13th MOA, by 42,250 feet of planted mangroves at the toe of the slopes of the bridge approaches, by placing litter within the swales created by the berms specified in the MOA. The berms, which had been constructed by a contractor without any apparent consistency of height or distance from the shoreline, protected the mangrove seedlings and propagules from strong wave action. If they were positioned correctly and erected to the proper height they were understood to play an important role in mangrove survival. Actual plantings of mangrove propagules that were done at four locations (Bahia Honda, Spanish Harbor, the Boca Chica Bridge Causeway, and Stock Island) satisfied the requirements for 16,600 linear feet. The plantings were done by inmates from the Florida State Road Prisons, under a contract between the DOT and the Florida Department of Corrections, at the specified one propagule every 5 feet. It was clear, upon completion of this phase, that the FDOT did not own sufficient right-of-way in the Keys to allow them to plant the additional 25,650 linear feet necessary to satisfy the terms of their agreement with the DER. They initiated a search for new sites on public land.

Road fill for the old Overseas Railroad and other development activities had cut off numerous inland lagoons on the Keys from their original access to ocean tides, resulting in increased salinity and consequent stress on associated living organisms. In their search for more sites for mangrove mitigation, FDOT personnel had identified two lagoons thus isolated as possible candidates: a 6.6 acre shallow bay-side lagoon on the west end of Bahia Honda, and a system of shallow semi-impounded interior lagoons on the Atlantic Ocean side of Boca Chica Key, roughly 2 miles by 0.6 miles in area.

A 20-foot swale was excavated through a mangrove ridge separating Bahia Honda from the Florida Bay in early April, 1981 (Fig. 6-3). Three monitoring surveys were done: one prior to the excavation, in February 1979, and two

Figure 6-3 Bahia Honda Lagoon. *Top*: circulation cut, 4/81; *Bottom*: circulation cut, 4/91. Reproduced with permission of Lewis Environmental Services, Inc. from *Wetlands Mitigation Evaluation Report Florida Keys Bridge Replacement*, 1994. Photos also appear in color insert.

subsequent ones, in August, 1981 and September, 1982 (Jordan, 1986). The surveys demonstrated the following results:

- A daily tidal regime was established that reduced the original hypersaline conditions.

- The blue-green algae that had dominated prior to the excavation disappeared and were replaced by *Batophora oerstedi*, an alga typical of local flushed lagoons. There were no changes in the benthic community.
- Mangrove growth and seed population were stimulated, particularly for black mangroves.
- Fish species present increased from two to seven and fish populations expanded by several orders of magnitude.

The 1993 survey recorded significant colonization and growth of volunteer red mangroves and numerous fish and wildlife species including reddish egrets, roseate spoonbills, and mangrove water snakes.

Encouraged by the observations at Bahia Honda, the FDOT initiated more extensive work on U.S. Navy property on Boca Chica Key. In early 1982, they installed three sets of 4-barrel 42 by 29-inch aluminum culverts about 0.5 mile apart under Old Boca Chica Road, connecting the south edge of the lagoons to the Atlantic Ocean (Fig. 6-4). The culverts were designed to produce an exchange of water and biota with the Atlantic, a flushing of the lagoons, and reduction in salinity. In addition to the culvert installation, the FDOT planted 80,000 red mangrove propagules, again with the use of prison labor, near culvert C-4, completing the work in August 1983, 18 months after the culverts were installed. Photographs of control plots taken in May 1985 showed healthy red mangrove seedlings that had developed, by 1993, into a flourishing red mangrove community (Fig. 6-5).

The 1993 survey of the lagoons connected to the culverts discovered new areas of seagrass where none had existed previously; an extensive new colonization and growth of both red and black mangroves; and a wide variety of fish, fiddler crabs, birds, and shorebirds. Although the extension of the range of mangroves at culvert 4, where 80,000 propagules were planted, was only slightly greater than that at culverts 3 and 2, where natural revegetation occurred, the 1994 report concluded that "Although pioneering mangroves are common in the interior here, the planting of mangroves resulted in considerably larger mangrove stands and larger trees" (Lewis et al., 1994). In total, almost 800 acres of mangroves had been enhanced and 33.37 acres of new mangroves had been created by the mitigation program (Table 6-4).

Approach Slope Revegetation

The issue of stabilization of the newly constructed bridge approach slopes arose out of concern for water quality degradation by erosion of the fine silts and sands of the construction materials, particularly during rainstorms. Sodding was considered to be prohibitively expensive and, as an alternative, the FDOT used a "mower clipping" technique, spreading cut mulch and seed from other grassy areas in the Keys. The DOT conducted preliminary tests in 1981 at the Lower Sugarloaf Channel bridge that were deemed by the DER to be successful and by the DOT to be ineffective (Lewis et al., 1994). However, the FDOT began

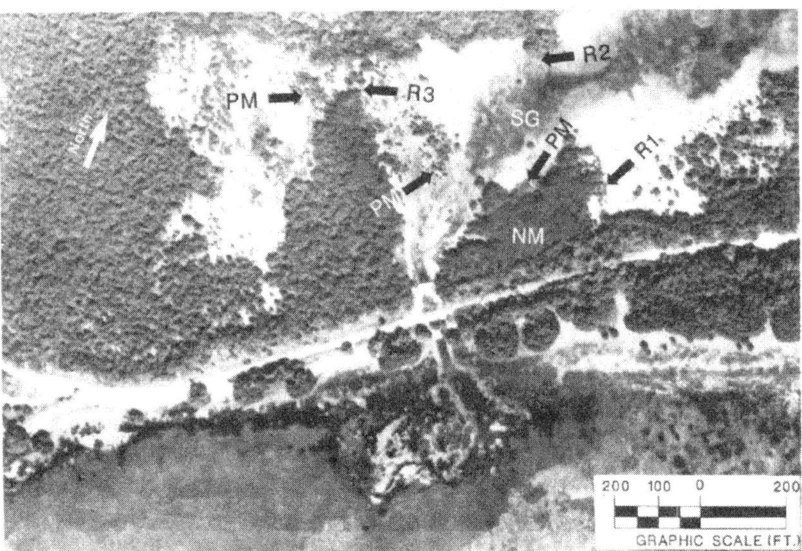

Figure 6-4 Boca Chica Lagoon site. *Top*: Vertical aerial photograph taken February 25, 1981, of the south side of the Boca Chica Lagoon at the location culvert C4. Reference points R1, R2, and R3 are also shown below. Arrow indicates future location of culvert (see below). *Bottom*: Vertical aerial photograph of the same site, taken February 17, 1991. Reference points are the same as above. R1 indicates a square planted mangrove plot. NM indicates new volunteer mangroves mixed with additional planted mangroves (PM) and SG new volunteer seagrass. Reproduced with permission of Lewis Environmental Services, Inc. from *Wetlands Mitigation Evaluation Report Florida Keys Bridge Replacement*, 1994.

Figure 6-5 Red mangroves at Boca Chica Lagoon. *Top*: Typical planted red mangroves at culvert C-4, Boca Chica Lagoon, 1985. *Bottom*: Typical volunteer and planted red mangroves at culvert C-4, Boca Chica Lagoon, 1993. Staff is 2 m tall. Reproduced with permission of Lewis Environmental Services, Inc. from *Wetlands Mitigation Evaluation Report Florida Keys Bridge Replacement*, 1994. Photos also appear in color insert.

to spread bagged mowed mulch on sites required by the DER, avoiding those slopes where least terns (*Sterna antillarum*) were discovered to be nesting. They completed the procedure at 14 bridges in the Lower Keys in 1982 and 1983, covering approximately 11,000 linear feet of new bridge approach slopes. The remaining bridges were exempted from treatment by the DER either because they were vegetating naturally or were being used for nesting by least

TABLE 6-4 Mitigation of Mangrove Losses

Locations of Mangrove Revegetation	Acres Lost	New Acres	Acres Restored
Bridge approach slopes	51.03	14.50	0
Bahia Honda	0	1.35	0
Boca Chica culvert 2	0	6.45	0
Boca Chica culvert 3	0	4.58	0
Boca Chica culvert 4	0	6.49	0
Boca Chica Lagoon	0	0	800
Total	51.03	33.37	800

terns. An example is depicted in the three photos of the Lower Sugarloaf Channel bridge approach slope (Fig. 6-6). The DER found the approach to be satisfactory and signed off on the DOT's slope revegetation efforts on July 15, 1983.

CONCLUSIONS

The ecological environments of the Florida Keys, with all their variety and complexity, are related in ways that are only partially understood. The importance of the known impacts of the Keys Bridge Replacement Project on these environments—the restriction of flows from the Florida Bay to the Atlantic Ocean; the increased turbidity resulting from the construction activities, the loss of seagrass and mangrove communities, and the denuding of bridge approach slopes; and the reduction of seagrass and mangrove habitat—are appreciated in the general sense without being totally understood in their specifics. The ultimate impact of the bridge replacement project is as undefinable today, therefore, as it was in the late 1970s when it was begun. In the short term, if 10 years can be considered short, the restoration of mangrove and seagrass communities to compensate for losses resulting from the Keys bridge replacements was very successful. Not only did it replace the lost ecological assets in the general vicinity of the project, but it also satisfied the demands of the natural resources agencies.

Some conclusions can be drawn that are related to that success:

- The negotiations initiated by the permit requirements resulted in avoidance and minimization of certain impacts that otherwise would have occurred: the bridges were replaced by other bridges rather than by filled causeways, bridge alignments were negotiated on a bridge-by-bridge basis to reduce specific impacts, appropriate spoil disposal sites were secured, and construction procedures that would minimize erosion were negotiated.

- On-site, in-kind compensation restoration activities were largely unsuccessful:
  - There was a net loss in turtle seagrass beds. The seagrass planting had a low survival rate and, where survival was observed, it was impossible to tell how much could be attributed to natural revegetation.
  - There was a net loss in shoreline mangroves because there was insufficient land available. Because original inventories recorded quantity but not quality, the relative values of the before and after mangrove communities cannot be assessed.
  - Bridge approach slope revegetation was successful quantitatively; again, the qualitative success is unknown because preconstruction inventories were not taken.
- Off-site compensatory restoration was extremely successful: when tidal circulation was reestablished at the Bahia Honda and Boca Chica interior lagoons, mangrove and seagrass communities were expanded well beyond the losses resulting from the project and attendant biotic communities expanded and flourished as well.

At least three functions were perceived as being affected by the project at its beginning: hydrologic flows through and between the islands by constructions caused by causeways; wildlife habitat by destruction of the fringe mangrove and seagrass meadows adjacent to the affected bridge sites; and water quality by destruction of the mangrove and seagrass communities, as well as by the loss of slope vegetation and construction activity. The hydrologic functions were largely preserved by eliminating causeways from project design. Habitat was definitely expanded in the Boca Chica lagoons where, if properly maintained, it can only be expected to improve. If offshore water quality has suffered as a result of the project, although there is no objective way of assessing whether it has, it would be an important impact that was not mitigated.

Could more have been done? Probably. It has been suggested that conditions of depth and thickness of underlying sediments could have been reestablished in the channel cuts where so much damage was done to seagrass meadows; earlier and more careful attention could have been paid to the slope revegetation; the design of the berms that facilitated the mangrove revegetation could

Figure 6-6 Approach slope revegetation, Lower Sugarloaf Channel. *Top*: Lower Sugarloaf Channel, Miami Gulf side, view west, 9/97. *Center*: Lower Sugarloaf Channel, erosion limiting revegetation, Miami Gulf side, view west, 5/93. *Bottom*: Lower Sugarloaf Channel, Miami Gulf side, view west, 4/82. The 4/82 and 5/93 photographs reproduced with permission of Lewis Environmental Services, Inc. from *Wetlands Mitigation Evaluation Report Florida Keys Bridge Replacement*, 1994 and the 9/97 photograph by N. Philippi. Photos also appear in color insert.

have been more precise—but this is all Monday morning quarterbacking. A lot less could have been done as well. Without the creativity and perseverance of some of FDOT personnel in accomplishing the interior lagoon flushing, the mitigation would have been a failure in every sense. Without any permitting process, the potential for environmental damage would have been unlimited—the tough negotiations among agency staffs forced compromises that would not have been accomplished even 10 years earlier.

That same combativeness, on the other hand, left a residue of bitterness that made truly adaptive mitigation difficult as the construction progressed and problems with the mitigation developed. The FDOT was apparently in no frame of mind to satisfy more than the letter of the agreement as the project developed. As one outsider observed, "getting the permits was a long and bloody process . . . the agencies kicked them [FDOT] around so that once they finally got them, they could have cared less." After 20 years, fortunately, the agencies have learned to work together much more smoothly.

ACKNOWLEDGMENTS

The Environmental Management Office of the FDOT was extremely helpful, Gary Evink in particular. Gary was the FDOT's state ecologist during the project, and today, as head of the Environmental Management office, he initiated and funded the important 1994 evaluation report. Charles Allen, who was the DOT permit administrator during the Bridge Replacement Program, and Dave Zeigler, who supervised all the mangrove and slope planting operations and participated in Operation Seagrass, were able to reconstruct large parts of the earliest period. Another valuable source of information was Roy (Robin) Lewis who, as founder of Mangrove Systems, was a principal in the Craig Key seagrass research and conducted Operation Seagrass and, as Lewis Environmental Services, did the 1994 evaluation report. Pat McNeese, of the Lewis Environmental Services office on Sutherland Key, enthusiastically tracked down all the old mitigation sites. Other important sources that took the time to stir up 20-year-old memories and dig into files were R. J. Helbling of the Marathon office of the Florida Department of Environmental Protection and a biologist for the Florida Department of Pollution Control in 1975; Hanes Johnson, now with the USEPA's Wetlands Section in Atlanta and a marine biologist for the DNR at the time of the project; Joseph Carroll, of Carroll and Associates in Vero Beach and a field supervisor with the FWS in Vero Beach when the project was underway; Shirley Stokes with the Army Corps of Engineers; Eric Hughes, with USEPA in Jacksonville; and Mark Thompson, with National Marine Fisheries now, and a biologist at USEPA at the time of the project. Kalani Cairns, a current FWS Biologist at Vero Beach, took the time and trouble to unearth old documents.

REFERENCES

Continental Shelf Associates, Inc., *Seagrass Revegetation Studies in Monroe County, Florida*, Florida Department of Transportation, Tallahassee, FL, 1982.

Douglas, M. S., *Florida: The Long Frontier*, Harper and Row, New York, 1967.

Jaap, W. C. and P. Hallock, "Coastal and Near Shore Communities: Coral Reefs," N. Phillips and K. Larson, editors, *Synthesis of Available Biological, Geological, Chemical, Socioeconomic, and Cultural Resource Information for the South Florida Area*, Continental Shelf Associates, Inc., Tampa, FL, 1990.

Jordan, W., *The Effects of Re-Establishing Tidal Circulation to a Mangrove Lagoon, Bahia Honda Key, Monroe County, Florida*, Florida Department of Transportation, Tallahassee, FL, 1986.

Kuyper, W., *Photographic Analysis of Bahia Honda Key*, Florida Department of Transportation, State Topographic Office, Tallahassee, FL, 1979.

Lewis, R. R., C. R. Kruer, S. F. Treat, and S. M. Morris, "Wetland Mitigation Evaluation Report Florida Keys Bridge Replacement," FL-ER-55-94, Florida Department of Transportation, Tallahassee, FL, 1994.

National Oceanic and Atmospheric Administration (NOAA), Florida Keys National Marine Sanctuary, Draft Management Plan/Environmental Impact Statement 1995, Volume II, Development of the Management Plan: Environmental Impact Statement, 1995.

Odum, W. E., C. C. McIvor, and T. J. Smith III, *The Ecology of the Mangroves of South Florida: A Community Profile*, Department of Environmental Sciences, University of Virginia, Charlottesville, VA, 1982.

Voss, G. L., *Coral Reefs of Florida*, Pineapple Press, Sarasota, FL, 1988.

CHAPTER 7

YAHARA RIVER MARSH

What was once rolling prairie capped by forests of oak and hickory interspersed, in the drainageways and low-lying landscapes, with wet prairies, sedge meadows, and marshes has been transformed over the past 150 years into a vibrant urban center in south-central Wisconsin (Fig. 7-1). In the 1850s, the natural landscape gave way to agricultural production and was replaced by the tilled field and grazed pasture. The upland prairies and forests were converted first, and the more-difficult-to-farm wetlands were converted last, if at all. Even after the complete settlement of the region, many wetlands remained. Typically, these recalcitrant landscapes survived along the drainage courses and around the numerous lakes in the area. But even these gave way to the development in and around Madison, Wisconsin's capital city, following World War II.

As development occurred, traffic congestion became an issue of great public concern. In the early 1960s, traffic congestion and accident rates were rapidly escalating along U.S. Highway 12, the transportation corridor traversing the southern edge of Madison (Fig. 7-2). Through this corridor flowed traffic from I 90 and 94 and U.S. 18 and 51 on the eastern edge of the city; U.S. 14, 18, and 151 on the southern edge; and U.S. 12 and 14 on the western edge. The congestion grew worse through the 1970s and 1980s as Madison and the surrounding communities expanded (Fig. 7-2). By the 1980s, traffic volumes far exceeded the design capacity of the existing corridor.

Engineers with the Wisconsin Department of Transportation (WDOT) grouped the solutions to the traffic problems into two alternatives: expansion and improvement of U.S. 12/18 or construction of a limited-access highway on a new alignment (referred to as South Madison Beltline) over the Yahara River and through the surrounding marsh. Although the solutions were easy to

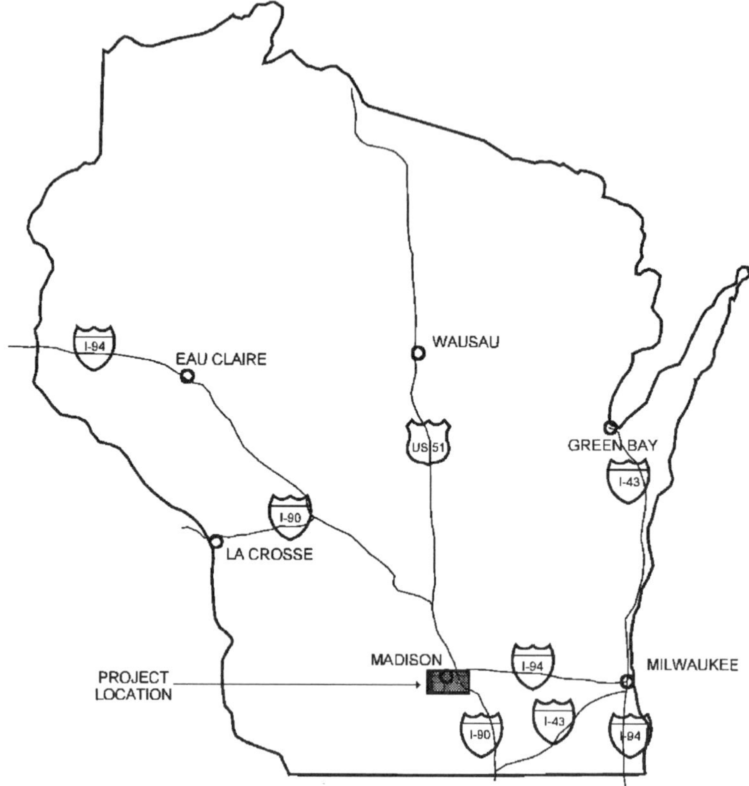

Figure 7-1 Project location.

characterize, 25 years and considerable public debate were required before a solution was approved and implemented.

The WDOT and the Wisconsin Department of Natural Resources (WDNR) engaged in a long debate over the environmental impacts and how they should be mitigated. The Public Intervenor, an agent of the Wisconsin Department of Justice, joined the fray, objecting to the new highway primarily on environmental grounds. The Wisconsin Wetlands Association was established to defend the integrity of the Yahara River and the enveloping marshlands, often referred to as the Upper Mud Lake Wetland but for this publication called the Yahara River Marsh.

In 1976, the citizens of Madison passed a referendum opposing a proposed new alignment (Boswell, undated). Concerns included an improved corridor accelerating urbanization; the loss of income and tax revenues along the existing corridor; the cost and inconvenience of construction; and the detrimental impacts to the Yahara River, the surrounding marshlands, and the dependent

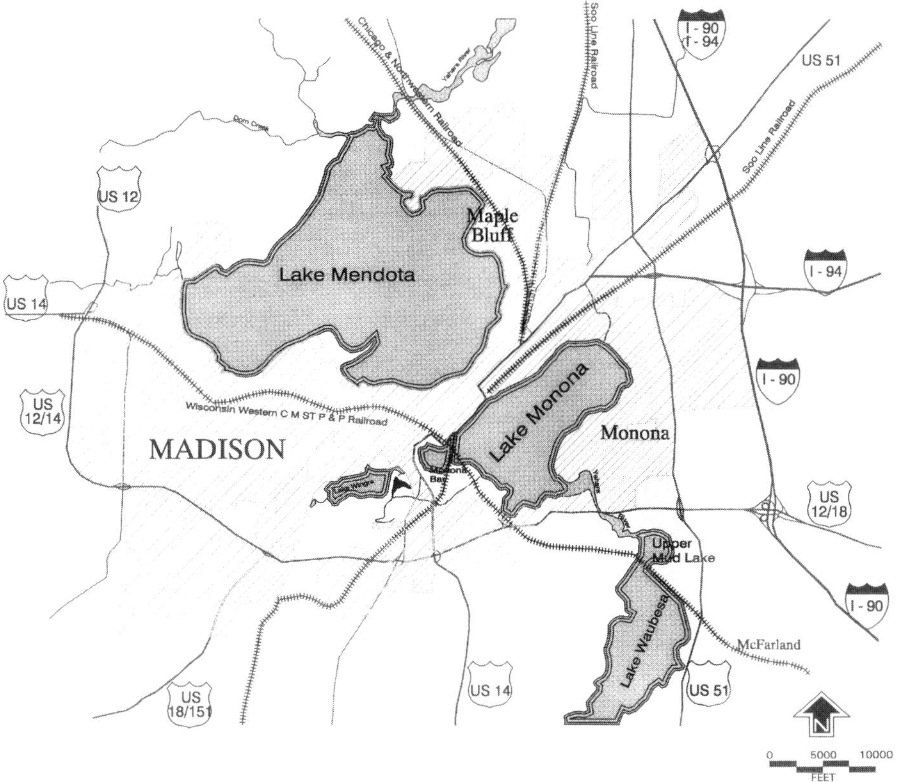

Figure 7-2 Metro-Madison.

wildlife (Fig. 7-3, aerial photographs before and after construction). The proponents for the project argued the economic need and the public safety issues while at the same time arguing that any wetland impacts could be mitigated. The opponents argued strongly against the project, citing urban sprawl and the loss of wildlife habitat. Some opponents claimed that the entire wetland complex along the Yahara River would be lost. The Wisconsin Wetlands Association seemed to mount the strongest opposing arguments. Their arguments were buttressed by the Public Intervenor. The collective environmental objections were enumerated by the Public Intervenor in a letter to the Division Administrator of the Federal Highway Administration (Falk, 1984):

> Much of Mud Lake and the Yahara River wetlands, which have existed for thousands of years, is zoned conservancy. In addition, they are located in Dane County's "E-way," an area designated by the Dane County Regional Planning Commission for preservation as open space because of their unique educational, environmental, aesthetic, and ecological values.

(*a*)

(*b*)

Figure 7-3 (*a*) Yahara River Marsh in 1973 and (*b*) after construction of South Madison Beltline in 1993.

> The net effect of the relocated beltline alternative would be that the public loses 22 acres of wetlands directly through construction of roadbeds and bridge approaches, and is left with 1,000 acres of adjacent wetlands in floodplains of greatly diminished value. Because size is the determining factor in how much recreational impact an area can absorb, and many diverse species of wildlife and wildfowl it can support, and its capacity for sediment deposition and nutrient filtering, a relocated beltline would adversely impair wetland and floodplain far in excess of the acres which are directly destroyed by the road.
>
> In addition, the common tern is found in upper Mud Lake Marsh and is listed on [Wisconsin's] endangered species list. Also found in the area is the Blanding's turtle, a reptile on Wisconsin's threatened species list.
>
> Numerous other reasons exist why this 1,000-plus acre wetland area is unusually significant: "its significant wildlife uses; its present and potential recreational area (especially given its close proximity to the large metropolitan area); its scenic, open space, and aesthetic values; its unusually large size; and its watershed protection value." These wetlands buffer downstream lakes and water from the negative effects of spring floods, sludge lagoons and excess nutrients, and urban and agricultural runoff.

The WDOT responded to each of these issues. On a point by point basis, the following paraphrases the Department's positions (FHWA, 1984):

> The Metropolitan Madison E-way is a concept proposing a linear system of natural and manmade features. The framework of the concept consists of a series of public streets, walkways, railroad corridors, and other open spaces connecting many of the area's more prominent educational and environmental features. The proposed E-way lies within the jurisdictions of Dane County, the cities of Madison and Monona and Fitchburg, and the town of Blooming Grove. Only the city of Madison has officially adopted the concept. The South Madison Beltline, which is a part of the areawide regional transportation plan for Dane County, passes through the conservancy area and along the northern edge of the proposed E-way. The city of Monona's Comprehensive Outdoor Recreation Plan states that the area zoned as conservancy may be developed consistent with that zoning after the WDOT has acquired highway right-of-way for the South Madison Beltline.
>
> Although the preferred alternative causes 22 acres of wetland to be lost, 20 acres of wetland will be re-created, 5 acres enhanced, and the WDNR will be given ownership of these lands and an additional 97 acres of wetlands currently owned by the WDOT, totaling 122 acres. Thus short-term losses are considered offset by long-term gains in diversity, productivity, and protection of the adjacent wetlands.
>
> Coordination with the WDNR and the U.S. FWS has been completed on possible endangered species impacts. The results of these reviews covering two federally and two state listed species follows:
>
> American Peregrine Falcon—The records of the Department of Natural Resources indicate that the last breeding adult was seen in Wisconsin in 1964. Further, the scientists with the U.S. FWS concluded that neither alternative would have an effect on critical habitat of this species.

Kirkland's Warbler—Although Dane County is within the migration range of the species, the Department of Natural Resources records do not include any sightings in Wisconsin.

Common Tern—This species is listed as endangered in Wisconsin. The Department of Natural Resources reported in 1981 that there were only two colonies left, both in Ashland Harbor on Lake Superior. Although more than 100 pairs nested in lower Green Bay in 1979, there were no nestings in 1980 in the waters of Lake Michigan. Based on this information, the apparent single sighting of this species by the consultant who conducted the waterfowl inventory could have been due to misidentification or an uncommon appearance of the species during migration.

Blanding's Turtle—This species was included as a potential wetland habitant based on the U.S. Army Corps of Engineers' *Bio-resources Inventory for Wisconsin.* Because the species is listed as occurring in Dane County, the Department of Natural Resources was contacted to confirm actual locations relative to the South Beltline corridor. The Department was informed that the nearest sightings were in the wetland south of Lake Waubesa and further that the wetland adjacent to the Yahara River widespread [the area through which the proposed road was to traverse] is unlikely habitat due to the vegetation and moisture conditions.

Also included in the discussions were species listed in Wisconsin's Watch Category—the great blue heron, black duck, marsh hawk, and common flicker have been confirmed as inhabitants of the Yahara River Widespread, Upper Mud Lake and their adjacent wetlands. Any significant continued loss of habitat for these species could result in a change in their status from watch to "threatened or endangered." The bullfrog was included as a potential wetland inhabitant according to the Army Corps of Engineers' computer inventory. However, there have been no actual sightings in the wetland.

The WDOT concluded that relocating the Madison South Beltline across the marsh would not significantly affect land of publicly owned parks, recreation areas, or wildlife and waterfowl refuges.

As the debate progressed, the WDOT studied numerous alternatives, including those proposed by the objectors. Two alternatives survived (FHWA, 1984). Despite the greater wetland impacts, the new alignment alternative became the recommended project. The reasons were numerous: greater safety; the displacement of fewer homes and business; less economic loss during construction to the businesses along the existing alignment; reduced noise impacts; and fewer conflicts with parks, boat landings, and archeological sites. The most telling reason, however, was the shift in public support for the new alignment. At the public hearing on the Draft Environmental Impact Statement (FHWA, 1983), 60% of the individuals who testified favored the new alignment; only 12% favored improving the old one.

This shift in public opinion may have resulted from the last minute endorsement by the WDNR of a modified new alignment. Following the hearing on the Draft Environmental Impact Statement, the WDOT, working closely with the WDNR, the U.S. FWS, and the Army Corps of Engineers, reduced wetland conversions from 31 to 22 acres. This reduction was achieved by using diamond rather than cloverleaf interchanges, adjusting the highway alignment, reducing median widths, and making the bridge over the Yahara River longer and more narrow (FHWA, 1984).

The WDOT also proposed to mitigate wetland losses. For the first time in the history of the Department, wetlands were created or restored in greater value and area than those being destroyed. The mitigation proposal included construction of sediment ponds to intercept highway runoff before reaching extant wetlands; removal of historic fill and restoration of the underlying wetlands on the Department's property; acquisition and restoration of selected, disturbed wetlands; acquisition of additional wetlands; construction of open water areas for waterfowl habitat; and preservation, in perpetuity, of all acquired and restored wetlands.

The end result was to be 22 acres of wetlands converted to highway use, 20 acres of wetlands created, 5 acres enhanced, and 122 acres protected in perpetuity. These numbers and conditions convinced the WDNR to endorse the project. Although the agency had opposed the project for a number of years, when an agreement was finally reached between the two agencies, they worked together to accomplish the mitigation objectives. The WDOT sought outside assistance in designing the mitigation effort and it maintained close control over the construction and subsequent monitoring and management of the restored landscapes. The WDOT seemed to have won over the skeptics in the WDNR.

The WDOT's engineers were able to modify the design of the project without sacrificing safety or capacity. The roadway was shifted south to avoid wetlands, and bridges, rather than embankments, were used to traverse wetland areas. Wetland losses were reduced from 72 acres, related to the highway design of 1972, to 31 acres at the time of the Draft Environmental Impact Statement (FHWA, 1983), and finally, to 22 acres owing to the objections of the Department of Natural Resources, the Public Intervenor, the Wisconsin Wetlands Association, and others. In the final analysis, the objectors had a pronounced effect on the design of the project and on the nature of the environmental mitigation.

Many of the objections were well reasoned and constructive. After listing the reasons why the road should not be built through the Yahara River Marshes, the Wisconsin Wetlands Association went on to acknowledge to the Army Corps of Engineers (Roherty, 1985), that

> . . . this [public notice] is largely a formality and that in all likelihood [the permit] will be granted and the project will proceed as planned. In that case, the crucial aspect of this permitting process is the conditions which you attach to the permit. In this regard, we request that you demand guarantees of the following measures to help assure that the project be carried out in the least environmentally damaging manner possible and that the wetlands restoration and enhancement projects have the greatest potential for success:
>
> That the DOT should budget adequate funds for the staff and resources for the planning and design of the restoration and enhancement projects. Uni-

versity, state agency, and private sector professionals in the environmental protection field should be consulted.

The DOT should select construction methods that minimize disturbance of the area. Particular attention to erosion control and approach routes for heavy machinery is needed. Again outside professionals should be consulted for innovations in these areas.

Funds allocated for mitigation should be used for that purpose at the site. The purchase of wetlands elsewhere is not an acceptable alternative.

Field work for restoration and enhancement should be opened up for bidding separately from the bidding for road construction. While the road construction firm could perform the earth-moving work involving heavy machinery, the smaller scale work, such as planting, should be done by someone with appropriate training under the supervision of wetland scientists.

The DOT should budget adequate funds for the planning and implementation of long-term monitoring to determine if the objectives of the restoration and enhancement projects are met.

The entire process and its end product should be put on public display to educate the public so that others can learn from it. Once the project is completed, an interpretive center should be located on site explaining the principles, and the difficulties, of wetland restoration.

On May 8, 1984, the Federal Highway Administration approved the Final Environmental Impact Statement. The WDOT submitted a Section 404 permit application to the Corps on December 6, 1984. By today's standards, the application was quite sparse: no detailed grading, planting, erosion control, management, or monitoring plans were attached. The Corps issued the obligatory public notice on December 20, 1984. The application was approved and permit number 85-136-02 issued on March 8, 1985, granting permission to "discharge approximately 271,550 cubic yards of granular material and rock riprap in approximately 22 acres of wetlands and waterway to facilitate the construction of the South Madison Beltline highway in conjunction with the installation of a 2,600-foot long bridge and a 10-foot wide by 8-foot high box culvert."

In the conditions attached to the permit, no mention was made of the recommendations by the Wisconsin Wetlands Association or any other objector. The WDOT, however, chose to act on most of the suggestions even though they were not mandated to do so. The WDOT retained a wide range of experts to develop a detailed plan, set goals and objectives, oversee construction, and manage and monitor the finished landscape. In fact, the only recommendation from the Wisconsin Wetlands Association that the WDOT did not implement was the suggestion that separate contractors be used for earth-moving and planting. Employees of the WDOT now acknowledge that planting would have been more successful if these contracts had been separated.

Construction on the highway began in the summer of 1985; wetland mitigation began in the early fall of that year. The highway and wetland mitigation were completed in the fall of 1986. Three years of monitoring followed, although not required by the Corps.

ENVIRONMENTAL AND SOCIAL SETTINGS

Located in south-central Wisconsin, the project falls within the temperate zone of the continental United States. The climate is moderately warm and humid. Approximately 32 inches of precipitation falls annually. Of this amount, 14 inches, or 44%, occurs as snow or sleet. The wettest month is July and the driest month is December. The annual mean minimum temperature is −18°F and the maximum is 95°F. The mean temperature is 48°F. Summer months are characteristically warm, with August—the hottest month of the year—having a mean monthly temperature of 70°F. The coldest month is January, averaging 10°F. Given different seasonal distributions, not all of the precipitation falling on the region is returned to the atmosphere by evapotranspiration. Of the 32 inches of precipitation, approximately 6.6 inches, or 21%, leaves the region as streamflow.

The Yahara River conveys the vast majority of streamflow to and from the project area. After climate, the river is the most significant defining force of the affected riparian marshes. The depth, frequency, and duration of surface inundation of the marshes are a function of the river's stage. The river's stage also affects groundwater elevations and the rate and quantity of water moving into and out of groundwater storage. The nutrient balance is inextricably tied to the river and the underlying soils. In turn, the hydrologic and water quality characteristics of the river are defined by the physical and social character of its watershed.

The Yahara River rises on the northern edge of Dane County. It flows north into Columbia County for a short distance and then turns south approximately 18 miles from the project. The river generally flows from north to south–southeast. It is a tributary of the Rock River, joining this stream, near Indian Ford, Wisconsin, about 25 miles southeast of the project. Its watershed, upstream of the project, is shaped somewhat like a light bulb, narrowing between two glacial ridges as it passes through Madison and the project site. The north–south axis of the watershed is 22 miles and the east–west axis is 17 miles. The elevation of the upstream watershed boundary is 1,050 feet above mean sea level (fmsl); the junction with the Rock River is 780 fmsl. Accordingly, the mean slope of the river is 12 feet per mile.

In total, the Yahara River watershed encompasses more than 500 square miles, of which 370 square miles are tributary to the project. The topography of the watershed is characterized by glacial features and deposits. Approximately 20 miles west of the project site is Blue Monds, the highest elevation in Dane County, 1,489 fmsl. The glacial drift is approximately 80 feet in depth, made up of clays, silts, sands, and gravel. The surface soils vary from one landscape position to another. On steeper slopes Dodge and St. Charles soils are found, whereas Otter and Orion soils are present in the lower-lying areas in and along the drainageways.

Based on 65 years of data (U.S. Geological Survey, 1997), at McFarland, Wisconsin (2 miles downstream of the project), the yield of the Yahara River is 6.62 inches and the mean annual daily flow is 159 cubic feet per second (cfs). The highest recorded annual mean is 336 cfs and the lowest is 63.8 cfs. This low variation in flow, despite extensive urbanization in the lower portions of the watershed, is due in large part to extensive wetlands in the watershed and to numerous lakes, particularly those in the Madison metropolitan area: Mendota, Monona, Waubesa, and Kegonsa. The stage of the river through the project corridor is controlled by a dam on the outlet of Lake Waubesa, where the U.S. Geological Survey's gage is located. A structure on the outlet of Lake Mendota controls flows entering the project area, attenuating flows moving downstream. During the late spring and summer, the normal water level between Lakes Waubesa and Monona, bracketing the project, is maintained at 849.6 fmsl.

Because the project is only a short distance upstream of the dam, and the intervening topography is very flat, water elevations throughout the project area reflect those at the dam, with only minor variations. The broad, low-lying floodplain through the project area accommodates considerable flood storage; consequently, flooding in this reach of the river is not an issue. Given the geomorphic and hydrologic conditions of the site, it is ideally suited for wetlands, and wetlands existed long before settlement.

The water quality of the Yahara River is considered to be good (FHWA, 1984). The dissolved oxygen remains high (10 to 12 ppm) throughout the stream system, and the nitrogen (NO_3) concentrations are low to moderate (0.21 to 2.8 ppm), as are phosphorus (PO_4) and suspended solids concentrations (0.04 to 0.12 and 2.9 to 28 ppm, respectively). Still, the waters are considered eutrophic. The abundance of macrophytes and frequent algal blooms validate this conclusion. Despite the urbanization in the lower portion of the Yahara River and the agricultural activities in the upper portion of the river, the quality of the river is remarkably similar in both reaches. Credit for the good water quality in the river might be given to the abundance of wetlands and lakes distributed throughout the watershed.

Owing to the flat, natural grade and the dams, the velocity of the Yahara River is very slow. This obviates bank erosion and channel scouring, which, in turn, minimizes turbidity. The clear water promotes the propagation of submerged and emergent vegetation. Combined, the slow-moving water and ample plant growth afford substantial water quality benefits.

Groundwater movement in the large wetland complex west of the river is from the northwest to the southeast. In the western upland fringe, the groundwater occurs at 13 to 25 feet below the soil surface. As the surface elevation falls, the differential becomes less. Hydric soils, such as Houghton, begin to dominate as the hydraulic gradient flattens on the approach to the river. In the

(a)

Figure 2-1 The effects of log jams on Locust Creek in north central Missouri. *(a)* Log jam impounding water and creating habitat. *(b)* Channel section after removal of log jam. *(c)* Sediment released by removal of log jam smothering herbaceous layer in forested wetland. Photographs by Ken McCarty, Missouri Department of Natural Resources.

(b)

(c)

Figure 6-3 Bahia Honda Lagoon. *Top:* circulation cut, 4/81; *Bottom:* circulation cut, 4/91. Reproduced with permission of Lewis Environmental Services, Inc. from *Wetlands Mitigation Evaluation Report Florida Keys Bridge Replacement,* 1994.

Figure 6-5 Red mangroves at Boca Chica Lagoon. *Top:* Typical planted red mangroves at culvert C-4, Boca Chica Lagoon, 1985. *Bottom:* Typical volunteer and planted red mangroves at culvert C-4, Boca Chica Lagoon, 1993. Staff is 2 m tall. Reproduced with permission of Lewis Environmental Services, Inc. from *Wetlands Mitigation Evaluation Report Florida Keys Bridge Replacement,* 1994.

(a)

(b)

(c)

Figure 6-6 Approach slope revegetation, Lower Sugarloaf Channel. *(a)* Lower Sugarloaf Channel, Miami Gulf side, view west, 9/97. *(b)* Lower Sugarloaf Channel, erosion limiting revegetation, Miami Gulf side, view west, 5/93. *(c)* Lower Sugarloaf Channel, Miami Gulf side, view west, 4/82. The 4/82 and 5/93 photographs reproduced with permission of Lewis Environmental Services, Inc. from *Wetlands Mitigation Evaluation Report Florida Keys Bridge Replacement,* 1994 and the 9/97 photograph by N. Philippi.

Figure 7-9 Salvaged marsh soil being deposited and spread. Photographs by Elizabeth Day, Wisconsin Department of Transportation.

Figure 7-10 Restoration Area 4—high and shallow marsh and open water. Photograph by Elizabeth Day, Wisconsin Department of Transportation.

Figure 8-6 Constructed rock weir in abandoned railroad embankment. Photograph by Kristine Meiring, Colorado Department of Transportation.

Figure 8-8 Restored sedge meadow with shrub carr fringe. Photograph by Kristine Meiring, Colorado Department of Transportation.

Figure 9-9 Wider portion of the oxbow during the dry season with high water indicated by dried algae on shore vegetation. Photograph by David Kelley, Kelley and Associates Environmental Sciences, Inc.

Figure 9-10 Kachituli Oxbow on right bank of the leveed Sacramento River. Photograph by David Kelley, Kelley and Associates Environmental Sciences, Inc.

low-lying areas adjacent to the river, the groundwater regime is very stable, saturating surface soils. Groundwater quality is considered good.

Curtis (1959) divided Wisconsin into two floristic provinces, divided by a line running from southeast to northwest, just north of Dane County and the project area. In the southern province, he identified seven plant communities: mesic, xeric, and lowland forests, prairie, oak savanna, pine barrens, and sedge meadow. Except for pine barrens, each of these communities was present in Dane County prior to European settlement. The forests and prairies were the first plant communities to be altered by settlement. They served to provide building materials and to cultivate agricultural crops. The low-lying areas survived longer because of their saturated soil and the extent of the engineering works necessary to drain them. Some of the sedge meadows bordering the Yahara River and present on the project site endured despite the changes in the upland landscapes and river system (Fig. 7-3). Curtis (1959, p. 365) provides a very clear description of the sedge meadow community:

> The sedge meadow is here understood to be an open community of wet soils, where more than half the dominance is contributed by sedges rather than grasses. As such, it is closely related on soils of similar moisture to fens, bogs, and wet prairies, among other open groups, and to the shrub thickets and wet forests of the closed communities. Under wetter conditions it grades to cattail and reed marshes or other emergent aquatic groups. It usually occupies a very low position in the regional soil catenas. The ground may be flooded in the spring or after heavy summer rains but it typically lies just above the permanent water table. The soil is either a raw sedge peat or a muck produced by decomposition of such peat, and is frequently incorporated with mineral matter deposited by overwash from the surrounding uplands. Water is always plentifully present and never a limiting factor by its lack. Excess water, however, may induce difficult conditions for many plants because of the disturbed oxygen relationships. The sedge meadow soils are frequently in a reducing condition and may reach extreme conditions in this respect, with abduction of methane or other highly reduced "marsh grasses."

This is one of the principal communities that the WDOT intended to restore and create as mitigation for losses along the new Madison South Beltline right-of-way.

A good variety of wildlife use the marshes along the Yahara River today. Regardless of the disturbance by past and present human activities, numerous species of mammals, birds, reptiles, amphibians, fish, and macroinvertebrates have populated these landscapes in the past. The Corps compiled a list of potential wildlife (FHWA, 1984), which includes 19 species of mammals ranging from white-tailed deer to meadow vole and 84 species of birds including the pied-billed grebe, wood duck, barred owl, and red-shouldered hawk. The listed 36 species of reptiles, amphibians, and fish includes the central newt, green frog, and carp. The list was simply a tabulation of potential species. Several wildlife surveys that were conducted during the design phase of the project showed that many of the species were present.

Dane County was settled in the 1840s. Farming expanded quickly from the first settlement. Almost the entire county was under cultivation by the end of the 1850s. Still, there were significant wetlands that were not drained and the Yahara River channel remained unaffected owing to the wide margin of marshes along its course. As urbanization increased, particularly after World War II, the remaining marshes gave way to parking lots, roads, and other impervious surfaces.

Madison was incorporated in 1862 and became the capital of Wisconsin in that same year (Durrie, 1874). Although it served as an administrative center in its early years, it has also served as the market center for produce, grain, and dairy products. After World War II, Madison began to expand rapidly and, at the same time, smaller satellite communities began to develop. The population of Dane County, the county in which Madison and the project are located, grew from 131,000 people in 1940 to 290,000 in 1970. Put another way, the population more than doubled in 30 years. In the process, more acres of farmland were converted to urban land, and the population density increased from 109 persons per square mile to 241 (Fig. 7-4). Today, 393,000 people live in Dane County, resulting in a density of 327 persons per square mile. Still, the predominant land use is agricultural, representing 80% of the 769,000 acres of land in the county.

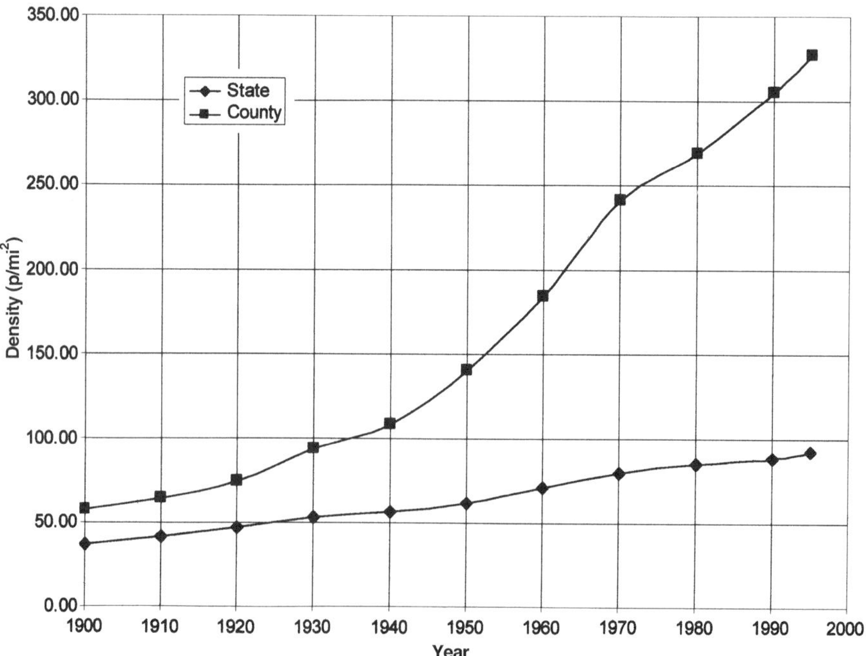

Figure 7-4 Population Densities of Dane County and Wisconsin.

With the arrival of European settlers, numerous plans were developed for the utilization of the Yahara River and its water resources. Navigation and hydropower were early interests, but throughout the public discussions of the plans, the recreational potential of the river and interconnected four lakes was not lost on the public or their political leaders. Several of the early proposals suggested development of resorts on the banks of the lakes. Today, the University of Wisconsin's Madison campus occupies the southern shoreline of Lake Mendota (also referred to as Fourth Lake). All four lakes (Mendota, Monona, Waubesa, and Kegonsa) were altered to one degree or another over the years. Each has a hydrologic control structure at its outlet and they are surrounded by development, receive runoff from highly urbanized watersheds, and are crossed by railroads and highways. Still, the lakes and their remaining wetland fringes offer high-quality recreational opportunities and, as such, provide great benefits to the residents and visitors of the metropolitan area. The proposed road would cross the Yahara River between the outlet of Lake Monona and the upper reaches of Lake Waubesa.

THE PROJECT AND ITS WETLAND IMPACTS

Based on traffic congestion and the incidence of traffic accidents, the WDOT concluded that a new, six-lane, limited-access roadway was needed to connect Interstate Highways 90 and 94 on the east side of Madison with U.S. Highway 12 on the west. Between these two highway corridors, a new limited-access route would pick up the all-important U.S. Highway 151 which traversed the Madison metropolitan area from southwest to northeast, as well as intercepting a number of other federal highways and major local roads. Traffic use along the existing east–west corridor had grown from 5,000 trips per day to 50,000 trips per day between the period when planning was first begun on the corridor in 1962 until the environmental impact statement was accepted in 1985.

The wetlands on the site were mainly deep and shallow marsh. The quality of these wetlands was not particularly high, but they were a part of a much larger complex of wetlands associated with the Yahara River and the adjoining lakes. They were valued for their wildlife habitat as well as for their plant community structure. Water quality was of some concern, but did not play a central role in the arguments for avoidance. Flood control was not an issue.

The project is within Dane County, traversing three municipal jurisdictions: the cities of Madison and Monona, and the town of Blooming Grove (Fig. 7-5). The six-lane freeway starts at John Nolen Drive on the west and extends to U.S. Highway 51 on the east, traversing 17,000 feet (3.22 miles). To reduce wetland impacts, the median strip was designed to be 24 feet wide, as opposed to the 66-foot width originally proposed. Excluding intersections, the pavement is approximately 50 feet wide in each direction and the typical right-of-way is 250 feet. In all, four intersections connect the freeway with other major highways and roads. The right-of-way required 121 acres of land. The bridge design

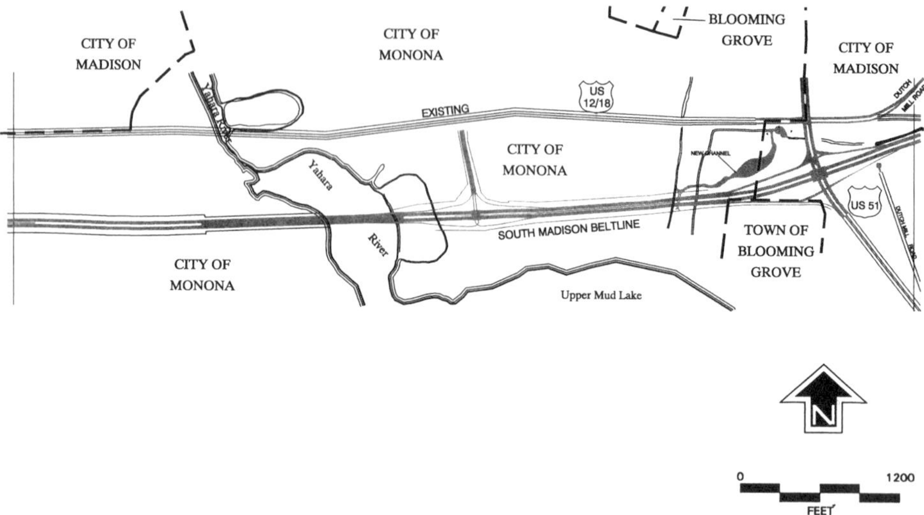

Figure 7-5 Political boundaries.

called for a 2,600-foot span—925 feet to cross the Yahara River and 1,675 feet to cross the associated marshes. Still, the area under the bridge deck was considered to be fill and was mitigated. Some 500,000 cubic yards of marsh soil (peat and muck) had to be removed to construct the road embankment. These materials, and others, were replaced by 1,500,000 cubic yards of borrow. Excavated marsh soil was stockpiled and used to spread in wetland mitigation areas.

The corridor along which the project was constructed varies in elevation from 860 fmsl on the east to 905 fmsl on the west. The low point is the Yahara River, with a normal water level of 845 fmsl. From east to west, the corridor dips toward the river, then gradually rises in the direction of Raywood Road. Passing over the western ridge, the corridor falls in the direction of John Nolen Road. Most of the affected wetlands are found in the west (right) bank of the river. Moving westward, after crossing the river, 2,300 feet of flat, low-lying marsh is traversed. This marsh is referred to as the Yahara River Marsh or Monona Marsh.

The soils in the corridor range from poorly drained (hydric), supporting marsh vegetation, to well drained, supporting upland forest. However, Houghton muck, a hydric soil, dominates the corridor (Fig. 7-6). Along the streambed of the Yahara River, silts and sands are found. Owing to the low velocity of the river, larger particle sizes are rarely encountered. As the elevation increases moving away from the river, drainage improves and forest soils begin to dominate. These include Dodge, St. Charles, Virgil, and Wacousta silt loams. The forest soils were first farmed and are now covered by urban development. The

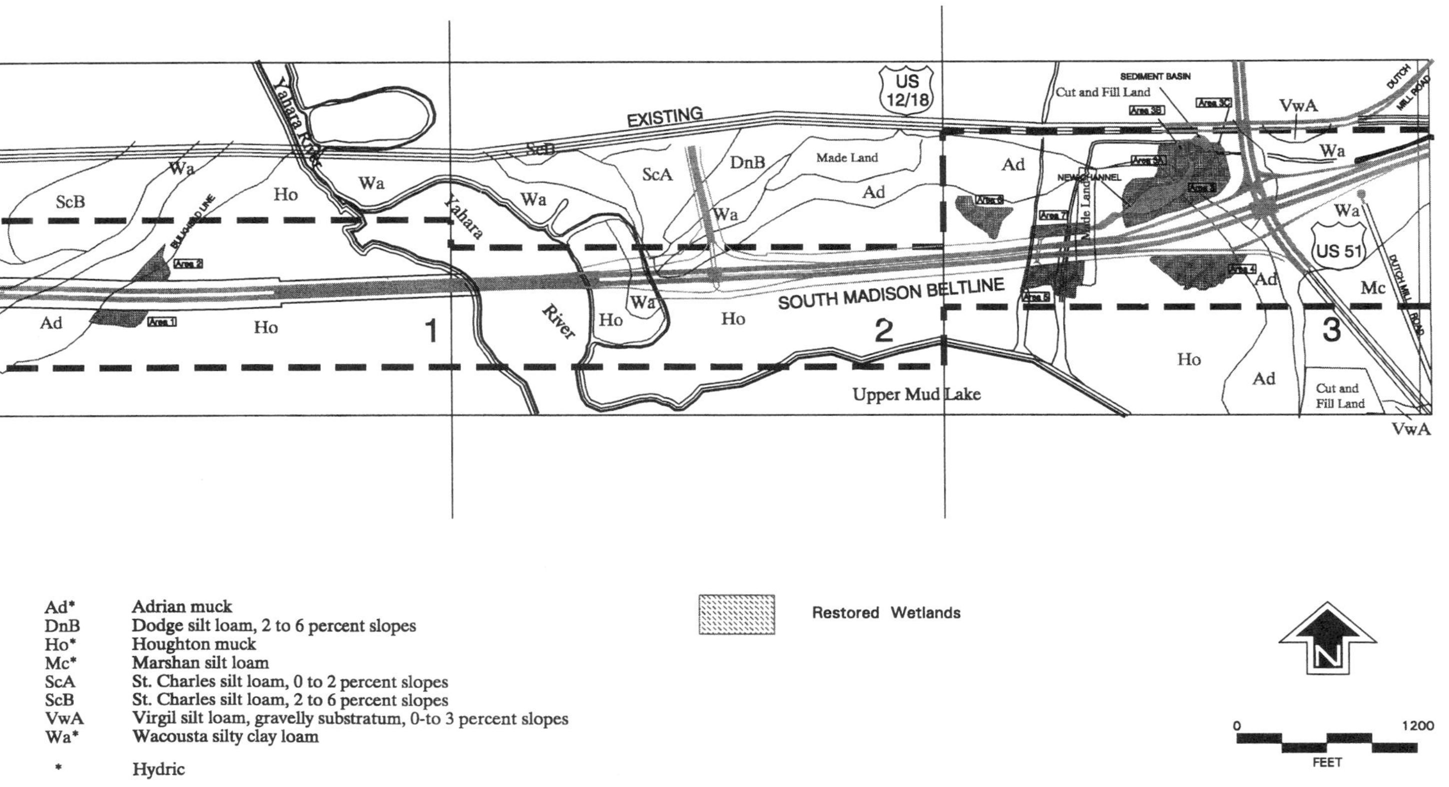

Figure 7-6 Soils map.

farming and construction difficulties associated with highly organic soils, such as Houghton muck, limited their development.

The vegetation of the corridor is dominated by bluejoint grass (*Calamagrostis canadensis*), sedges (*Carex* sp.), cattail (*Typha* sp.), reed canary grass (*Phalaris arundinacea*) and, in the higher elevations, some woody vegetation such as oak (*Quercus* sp.), box elder (*Acer negundo*), and Eastern cottonwood (*Populus deltoides*) (Fig. 7-7). Following the cessation of farming or other uses, the plant communities were left to develop on their own. Fire or other management techniques were not applied.

The wetlands of the Yahara River Marsh are characterized by Bedford et al. (1974) in accordance with predominant plant groupings:

- Deep marsh, mostly stands of narrow leaf cattail.
- Shallow marsh, stands of various plants, alone or mixed.
- Sedge meadow, with sedges and bluejoint grass.
- Dry sedge meadow with forbs.
- Shrubs.

Based on a field survey of the marshes, Bedford et al. (1974) reported that

> The west side marsh [west or right bank of the Yahara River] is a large peat bed. . . . The open water where the river widens is noted as a place to watch migrating ducks in spring. . . . The east side wetland is mostly sedge meadow on peat, and is drier than the west side. There are several small ditches, a large area where shrubs and some trees are invading, and a large reed canary grass area. Along the river edge is a wider, narrow-leaf cattail strip than is found on the west side of the river. There appears to be considerable woodcock habitat here: breeding snipe and woodcock were noted. Much of this area offers fair isolation from human disturbance.
>
> On the west side, dominant plants in separate and mixed stands are cattail, burreed, lake sedge, bluejoint grass, various wetland forbs, shrub willows, and redosier dogwood. Some bog birch, a stand of cordgrass, and a few stands of giant reed were noted. Numerous muskrat houses and bullfrogs were seen.
>
> An interesting plant community of sedges on floating mat was found on the far west side.
>
> Along the edge of the river much of the narrow leaf cattail strip is on a built up organic material which is dry enough for jewel weed to invade.
>
> The big wooded island on the west side contains a mixture of burr and black oak, black cherry, basswood, yellow bud hickory, and aspen. Probably both prairie relics and forest wildflowers exist there because summer access by people and cattle is unusually difficult.
>
> On the east side of the river are large areas of sedge and bluejoint. Near the river are stands of narrow leaf cattail. Willows, red-osier dogwood, and aspen are heavily invading a ten-acre plot as well as surrounding areas.

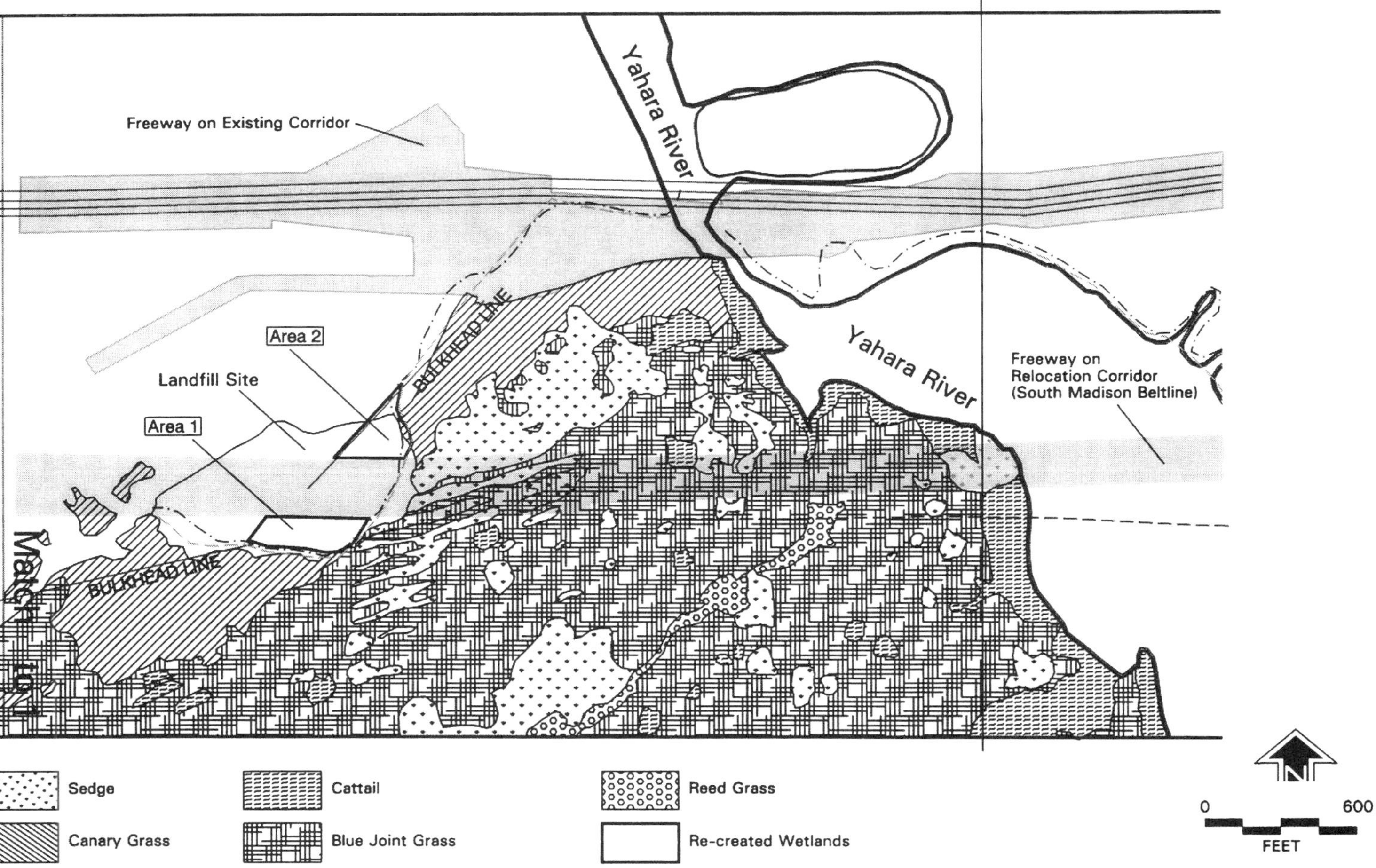

Figure 7-7 Plant communities.

By and large, cattail, bluejoint grass, common lake sedge (*Carex lacustris*), red-osier dogwood (*Cornus stolonifera*), and reed canary grass cover most of the wetland area. A similar distribution of plant communities existed along the highway corridor. Of the 22 acres of wetlands scheduled to be lost to the freeway, 10% were dominated by sedges, 40% by cattail, and 46% by reed canary grass.

The wetlands that were traversed by the highway were by no means pristine. In 1940, a dam was built across the Yahara River at the outlet of Lake Waubesa, downstream of the project, which altered the hydrology of the entire area. The plant communities adapted and new wetlands emerged. Subsequently, urbanization altered the hydrology of the local drainage areas, and the edges around the marshes were filled for commercial and transportation purposes and waste disposal. Several commercial enterprises had pushed materials out into the marsh in order to make more room for storage adjacent to their properties. Foundry sand was dumped as waste in the marsh as well as construction debris and other rubble.

MITIGATION IMPLEMENTATION AND RESULTS

As it became apparent that the WDOT's proposed South Beltline would be supported politically, the arguments began to shift from opposition to the road to how best to mitigate the environmental impacts, particularly the effects on wetlands. In this determination, scientists from a number of agencies and the University of Wisconsin participated.

The FWS sought the establishment of open water marshes to be used for waterfowl and to provide year-round fish habitat. Scientists from the University of Wisconsin were particularly concerned with plant diversity and focused on the restoration of sedge communities within mitigation areas. The WDOT hired several graduates of the University of Wisconsin to work on the mitigation plan and to coordinate the Department's efforts with the WDNR, the FWS, and the Corps. In the end, a plan was developed that met the requirements of these agencies.

The goal of the mitigation plan was to create wildlife habitat (principally for fish, waterfowl, and those animals associated with sedge meadows), not just wetlands. This goal, which was not explicitly stated, was assumed to be achieved given the creation of high-quality wetlands. The definition of high-quality wetlands was taken from Bedford et al. (1974), and includes the following characteristics:

- High water quality (including lack of silt or excess nutrient input).
- Natural water level cycle.
- Plant and animal species diversity.
- Structural diversity (i.e., mix of tall and short plants, open water and marshes).

- Edge gradation (created by gradual slopes).
- Lack of non-native or exotic species.

The plan went on to call for specific measures to offset the wetland losses:

- Recreating 20 acres of wetlands and enhancing an additional 5 acres of existing wetlands.
- Creating a sediment pond to trap contaminants and improve water quality of storm water runoff before entering Upper Mud Lake.
- Acquiring all privately owned wetlands between Upper Mud Lake and the existing Broadway transportation corridor—these lands would be retained in public ownership for long-term protection.

The aggregate amount of land in the mitigation program was 122 acres.

The WDOT committed to meeting the goals and objectives by the acquisition of land, restoration of wetlands that had been destroyed or modified in the recent past, and conveyance of the property along with other sections of the Yahara Marsh to public ownership. The City of Monona became the management agency.

The wetland restoration component of the mitigation plan consisted of 10 separate projects, numbered 1 through 7 and 3a, 3b, and 3c (Fig. 7-8a through 7-8d). They were distributed along the highway corridor near the affected wetlands and connected by the marsh matrix in which they were placed. The area of these projects totaled approximately 20 acres (7.5 hectares), which, as it turned out, exceeded the area of filled wetlands. After the Environmental Impact Statement was prepared, it was determined that only 18.3 acres of wetlands were displaced, as opposed to the reported 22 acres (WDOT, 1984).

Of the restored area, approximately 2.7 acres (14% of the total) were reclaimed from a foundry sand dump. Fill from a miniature golf course was removed to create 9.9 acres (50% of the total), an old auto salvage yard was reclaimed to create 3.3 acres (17%), and demolition debris was removed to restore 2.6 acres (13% of the total). The remaining 1.5 acres (6%) were created from weed-infested, abandoned agricultural fields and highway embankments.

For each site, design objectives were established (Table 7-1). These objectives covered plant species and diversity and habitat structure, ranging from shallow marsh to open water. The plant species for each of the design objectives and the two planting phases are given in Tables 7-2 through 7-4.

Three vegetation establishment methods were considered in the planning phase: natural recolonization, planting, and transplanting marsh sod. Exclusive use of natural recolonization was discounted because of the possibility of large-scale invasion by aggressive, non-native plants or the creation of cattail monocultures. Planting, on the other hand, was viewed as possibly limiting plant diversity owing to the small number of wetland plants commercially available. The contractor might have been asked to hand collect the desired species, but the environmental impact of this activity, given the large number of plants

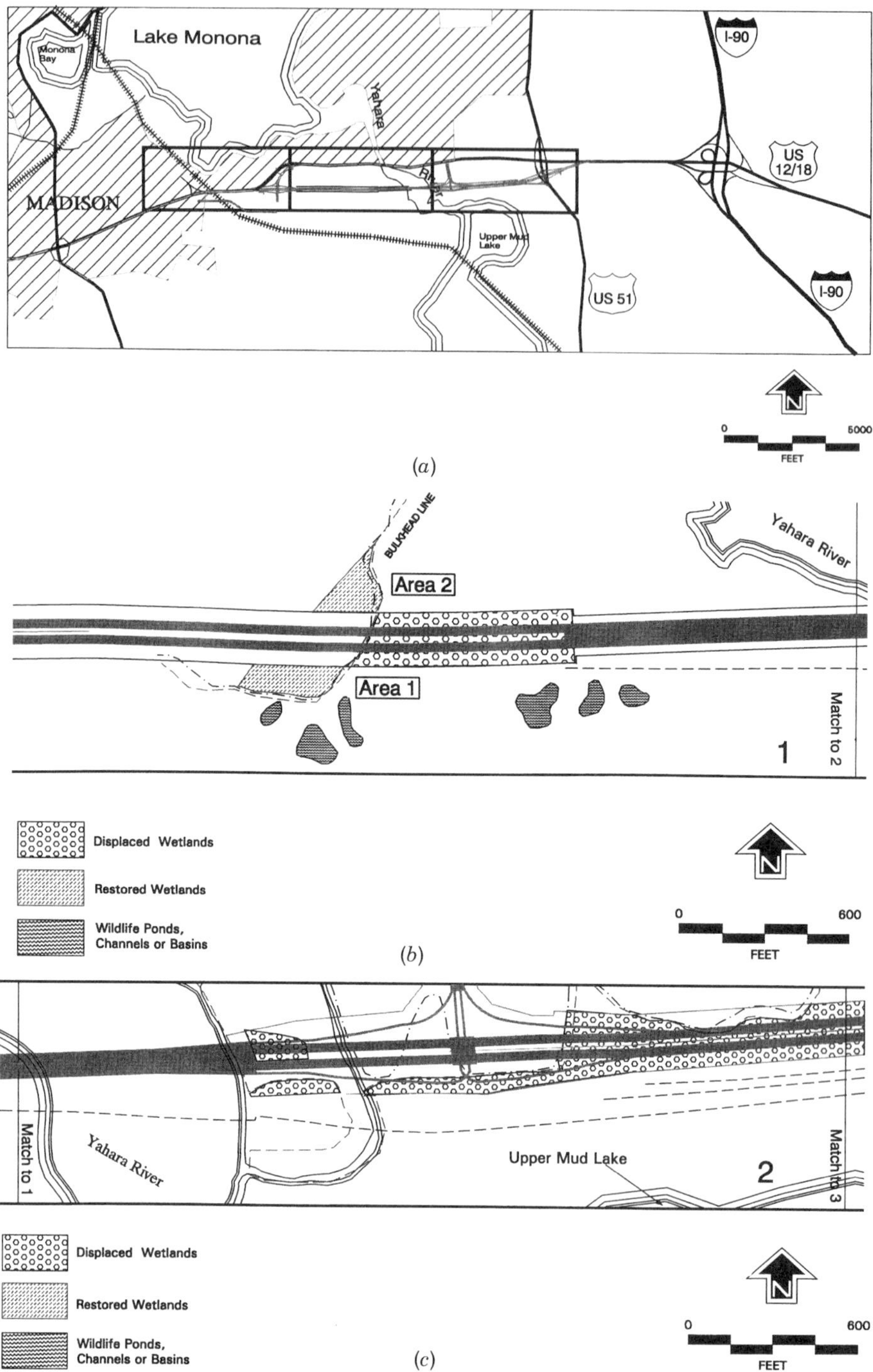

Figure 7-8 Wetland fill and mitigation site. (*a*) Overall site map. (*b–d*) Detailed maps showing sites 1–7. *Illustration continued on opposite page*

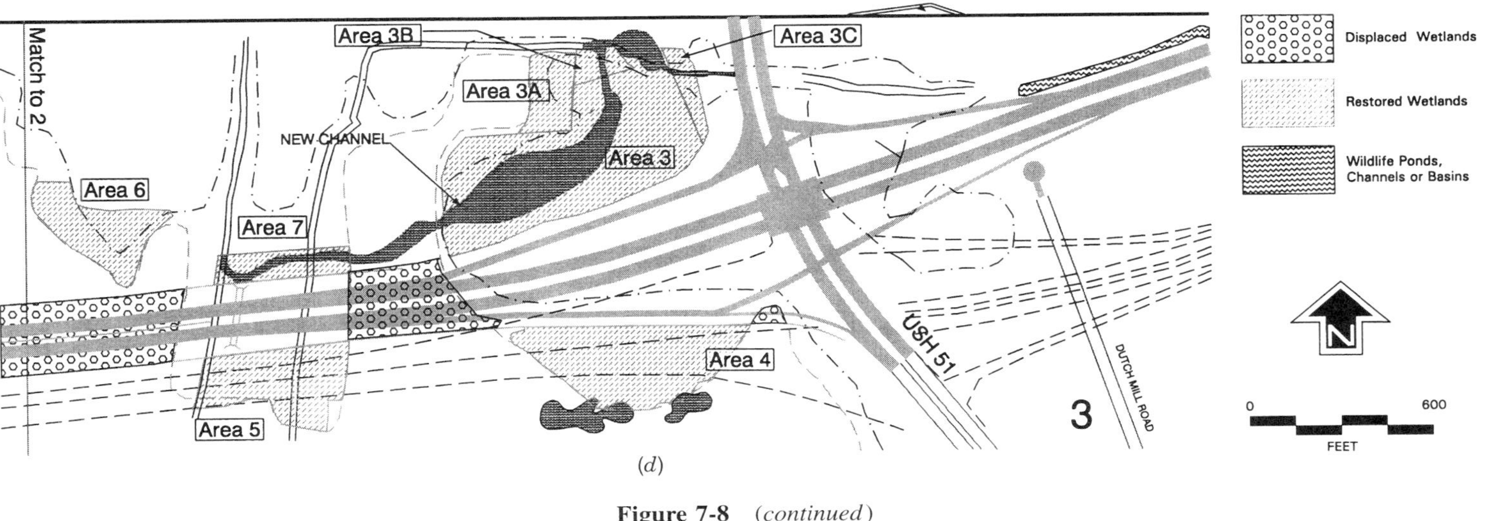

(*d*)

Figure 7-8 (*continued*)

TABLE 7-1 Restoration Areas and Design Objectives

Wetland Number	Acres	Hectares	Premitigation Status	Design Objectives
1	2.2	0.9	Foundry sand dump	Shallow marsh
2	0.5	0.2	Foundry sand dump	Shallow marsh
3	7.6	3.1	Miniature golf course fill	Shallow and deep marsh and open water
3a	1.0	0.4	Agricultural land	High and shallow marsh
3b	0.0	0.0	Agricultural land	Detention basin
3c	0.5	0.2	Highway embankment	High and shallow marsh
4	2.3	0.9	Miniature golf course fill	High and shallow marsh and open water
5	1.6	0.6	Auto salvage yard fill	High, shallow and deep marsh
6	2.6	1.1	Demolition debris fill	High, shallow and deep marsh
7	1.7	0.7	Auto salvage yard fill	Deep marsh
Total	20.0	7.5		
			Road opened in December 1988	

TABLE 7-2 Wetland Species Planted in Phase 1 Mitigation

Common Name	Scientific Name
Zone 1: Shallow Marsh (0.5 to 1.0 ft of water)	
Bur reed	*Sparganium eurycarpum*
Duck potato	*Sagittaria latifolia*
Marsh smartweed	*Polygonum muhlenbergii*
Pickerelweed	*Pontederia cordata*
River bulrush	*Scirpus fluviatilis*
Sweet flag	*Acorus calamus*
Zone 2: Deep Marsh (1 to 2 ft of water)	
Deepwater duck potato	*Sagittaria rigida*
Hardstem bulrush	*Scirpus acutus*
Sage pondweed	*Potamogeton pectinatus*
White water lily	*Nymphaea tuberosa*
Wild celery	*Vallisneria spiralis*
Zone 3: High Marsh (0 to 1.5 ft above water table)	
Common reed grass	*Phragmites communis*
Prairie cordgrass	*Spartina pectinata*

Source: Crabtree, 1992.

needed, was thought to be significant. Spreading the upper layer of excavated marsh soil on restoration sites having similar hydrologic conditions was expected to introduce a rich seed bank and a large quantity of native root stock at the new site. The macro and micro flora and fauna of the soil environment also were assumed to be moved with the soil. It was concluded that a solid layer of excavated marsh soil would discourage growth from the underlying seed bank, which was predominantly reed canary grass. Germination experiments showed that the seeds in the marsh soils were still viable.

Grading plans were based on one principle: gradual slopes. Natural wetlands, the scientists argued, are quite flat. Most of the slopes in the restoration sites were less than 2% (Owen et al., 1989). Elevation and slope dictate the water depth at a given point, which, in turn, dictates the plant species that will inhabit a given location. Gradual slopes encourage stable plant communities, which are adapted to seasonal water level fluctuations. Their argument continued: completely flat wetlands would be severely lacking in diversity.

Wherever possible, contour grades were drawn with no more than a 1:50 slope. This meant that the larger wetland restoration sites could include a range of wetland types, from sedge meadows to deep marsh. The drier wetland types were planned for smaller sites. Slopes as steep as 1:10 were used for up-

TABLE 7-3 Wetland Species Seeded in Phase 2 Mitigation[a]

Species	Common Name
	Annuals
Impatiens biflora	Jewelweed
Polgonum lapathifolium	Willow weed
Polygonum pensylvanicum	Smartweed
	Perennials
Angelica atropurpurea	Angelica
Asclepias incarnata	Marsh milkweed
Aster novae-angliae	New England aster
Calamagrostis canadensis	Bluejoint grass
Eupatorium maculatum	Joe-pye weed
Eupatorium perfoliatum	Boneset
Polygonum coccineum	Water smartweed
Polygonum hydropiperoides	Mild water pepper
Pycnanthemum virginianum	Mountain mint
Rumex orbiculatus	Marsh dock
Scirpus cyperinus	Wool grass
Thalictrum dasycarpum	Meadow rue

Source: Crabtree, 1992.

[a]Seeding rate = 2 lb/acre; seeded in high and medium marsh zones.

TABLE 7-4 Wetland Species Planted in Phase 2 Mitigation

Species	Common Name	Zones[a]
Carex lacustris	Lake sedge	H, M
Carex stricta	Tussock sedge	H, M
Carex hystericina	Sedge	H, M
Calamagrostis canadensis	Bluejoint grass	H, M
Spartina pectinata	Prairie cord grass	H
Iris virginica shrevei	Blue flag iris	H
Sparganium eurycarpum	Bur reed	M, L
Scirpus fluviatilis	River bulrush	M, L
Scirpus validus creber	Soft stem bulrush	M, L
Sagittaria latifolia	Duck potato	M, L
Sagittaria rigida	Arrowhead	L
Scirpus acutus	Hard stem bulrush	L

Source: Crabtree, 1992.

[a]H = high marsh, M = medium or shallow marsh, L = low or deep marsh.

gradient wetland restoration zones to reduce the area subject to invasion by upland pioneer species. Deep-water restoration areas and wildlife ponds in existing wetlands were also designed with steep slopes to discourage vegetation growth and conserve space. Open water areas were given natural-looking shapes in order to enhance aesthetics. Irregular shorelines also create more habitat for waterfowl.

The WDOT undertook the mitigation work with the direct involvement of the team of scientists (university, contractors, and staff) responsible for preparing the mitigation plan. The excavation and planting contracts were part of the general contract issued for construction of the roadway. This contractual arrangement ignored a suggestion made by the Wisconsin Wetlands Association that a separate contract be issued for the mitigation work and, in the end, it proved to be troublesome—the general contractor had little experience with wetland restoration. Construction engineers along with the design team oversaw the restoration activities.

To minimize impacts on the marsh, the bridge over the Yahara River and road embankment through the adjoining wetlands were constructed from the bridge deck and road surface. The embankments, foundation structures, support columns, and bridge deck were constructed without entering the marsh. Equipment and materials were advanced along the completed bridge and road sections.

Early in the construction planning phase, an investigation was launched into the possibility of regulating water levels of the Yahara River, which would in turn affect water levels in the mitigation sites. The idea was to use the dam and control structure on the outlet of Lake Waubesa, 2.3 miles downstream of the restoration sites. This would afford a very quick and easy way of achieving the desired hydrologic effects. In fact, if properly managed, some of the excavation might have been avoided—the water level could have been raised rather than the land lowered, but this would have affected other areas as well. The WDNR, among others, did not look favorably upon the idea because of the interference with recreational and wildlife uses. The WDOT was also concerned about the long-term management requirements of manipulating water levels. Consequently, this idea was abandoned.

The idea of using excavated marsh soil, largely Houghton muck, proved successful. The material was removed from the highway right-of-way and spread on selected restoration sites (Fig. 7-9). When road construction preceded wetland restoration, the topsoil was stockpiled. After several trials of distributing the excavated marsh soil, the best technique involved excavating a small area of the filled material from the restoration site, approximately 6 inches below the final grade, and then distributing the marsh soil. The excavating equipment was then moved back, and more fill material was removed and marsh soil distributed. Some work was done during the winter when the ground was frozen, and this facilitated the movement and final grading of the marsh soil.

A range of construction techniques was employed. Of the 10 restoration sites, seven required the removal of unwanted material: foundry sand, uplands

Figure 7-9 Salvaged marsh soil being deposited and spread. Photographs by Elizabeth Day, Wisconsin Department of Transportation. Photos also appear in color insert.

soil fill, or debris. In two cases, 3a and 3c, in sites that were formerly agricultural land, only shallow grading was required. The grading was used to remove the top few inches of soil in order to rid the site of reed canary grass. The excavated materials from sites 4 and 7 were disposed in the road embankment when suitable for such construction. The other materials were hauled off site and disposed in landfills.

Sites 1, 2, 3, and 4 were excavated and marsh soil spread during the first year of construction, September 1985 through September of 1986. The remaining sites were constructed after the road embankment was complete and no marsh soil was available. Consequently, sites 3a, 3b, 3c, 5, 6, and 7 required a different planting scheme. For those sites receiving the marsh soil, it was assumed that the seed and plant materials were in the soil and that they were viable, which was generally the case. For the other areas, because they were not excavated to the depth of the old, underlying marsh soils, the remaining surface soils were believed to contain only weeds or no viable plant materials. Consequently, these sites were planted with rootstock and seeded with a cover crop.

Some difficulties arose during restoration work. Wrong species of plants were delivered in some cases and in others the rootstock was unacceptable. Elevations had to be closely monitored. However, as construction proceeded, because of the presence of trained scientists, these problems were readily overcome.

In the end, the mitigation landscapes were established and planted. The desired results now depended on the ensuing hydrologic conditions and on the physical and biological interactions among the new soil surfaces and the hydrology, plants, and wildlife. Work started on the restoration sites in September of 1985 and was completed by May of 1988. The highway was open for traffic in December, 1988.

Although monitoring was not required by the Corps, the WDOT did monitor selected restoration sites for 2 years (Jackson, 1990). Also, studies were undertaken by graduate students and faculty of the University of Wisconsin (Owen et al., 1989; Ashworth, 1992) and independent contractors (Crabtree, 1990). These studies focused on selected restoration areas and compared them to referenced sites located in the Yahara Marsh.

At the end of the first phase of restoration (completion of sites 1, 2, 3, and 4), the work was evaluated before proceeding with the second phase (restoration of sites 3a, 3b, 3c, 5, 6, and 7). The idea was that some knowledge would be gained in the first phase that could increase the chances of success of the second phase. Following the initial monitoring effort, the mitigation strategy was changed: cover crop usage, expanded plant list, and mulch application. No further changes in strategy occurred although monitoring continued. These later monitoring efforts resulted in comparisons between restoration and reference sites and in identification of design and construction flaws. Still, these reports provide useful insights to future restoration activities.

As reported by Day (1986), wetland vegetation began to emerge from the spread marsh soil early in April, 1986. Sedges were first, followed by cattail, then bur reed and arrowhead. These plants and others prospered through the summer and by September they had reached heights of 3 to 6 feet. Still, Day expressed concern for the fragile nature of these newly developing ecosystems.

Over two growing seasons, the plant community in the restored areas continued to develop, but somewhat differently from that found in reference areas. For example, sedges and bluejoint grass were notably absent in Area 3 whereas they were frequently found in reference areas, but the presence of cattail was similar for both reference and mitigation areas. Further, the reference areas had fewer non-native species than the mitigation areas. Day (1986) attributed some of these differences in Area 3 to

> . . . a "younger'' community than the source wetland in terms of ecological succession. Even though all the rootstock from the source wetland was present in the salvaged marsh surface (SMS), the growing conditions in Area 3 (i.e. exposed mudflats) favored the pioneer species present in the seed bank. The straw mulch specified for these areas was originally intended to shade the exposed mudflats and favor growth from rootstock. Unfortunately, data are insufficient to evaluate differences between mulched and unmulched areas. In any case, the mulch layer specified was probably too thin to duplicate the influence of a normal detritus layer along with the standing dead stems found in a well established marsh. Restoration Area 3 has only been through one growing season. Its character could change significantly over the next several years.
>
> The second explanation is that the hydrologic regime of Area 3's SMS zone is quite different from that of the source wetland. The average substrate elevation in the sampled portion of Area 3 is 844.6. The average elevation of the source wetland was 845.3. The difference in elevation, i.e. deeper water, seems to have discouraged colonization by sedges and bluejoint grass. This was anticipated but not to the extent that it occurred. The actual range of substrate elevations is approximately one-half foot lower than the intended range. This is probably due to the difficulties involved with detailed grading underwater.

As new as the plant communities were, and as fragile as they may have been, they attracted a wide range of wildlife within their first year of existence. Mallards, pied-billed grebes, blue-winged teal, and the ever-present Canada geese occupied the open water areas. They used the habitat for breeding and foraging. A large population of migratory shorebirds, such as greater yellowlegs and dowitchers, found the mud flats attractive habitat. Green and great blue herons, both wading birds, were seen in and around the restoration sites during the summer. Chorus and pickerel frogs, turtles, and small forage fish moved into the newly created landscapes from surrounding habitat. Muskrats colonized Areas 2, 3, and 4, causing some concern because of their habit of eating young plant shoots. Because trapping was ruled out, their presence was tolerated.

Day (1986) noted several problems with the execution of the construction plan:

- If grading of the marsh soil resulted in high spots, often these areas were colonized by woody species and later by undesirable weeds.
- In some cases, the spread marsh soil resulted in a healthy, viable plant community whereas, in the same location, plantings failed, seemingly without explanation.
- Erosion and a deposition of sediment materials over previously graded and planted landscapes, in isolated cases, led to colonization by undesirable species.
- In some cases, a grid of planting on 6-foot centers left open areas that were quickly colonized by weedy species.

In the early stages (i.e., the first 2 years), the monitoring program revealed the failure of certain plants to propagate, requiring a second planting. This was done in 1992, in the mitigation Areas 5, 7, and 9. In other cases, the desired plant communities failed to develop but hydrophytes prospered. Unfortunately, a drought occurring in the second year contributed to the rapid expansion of less desirable wetland plants, such as cattail and the non-native canary grass.

Management included some manipulation of hydrology. Irrigation was used at the early stages of reestablishing plant communities. In large, however, the hydrology of the marshes was left up to the existing climatic conditions and the normal effects of the dam on Lake Waubesa. Fire was used but only to a very limited extent. Grazing by muskrats and other herbivores was not controlled or limited. Attempts, however, were made to prevent geese from consuming the early plantings.

In the end, wetlands were established in the mitigation areas but they did not meet the design criteria in all cases (Fig. 7-10). Plant communities differ from the design criteria and the presence of reed canary grass and other invasive, non-native species plays a much bigger role in the marshes today than anticipated by the designers in 1988. Still, the restored wetlands provide wildlife habitat, flood control, and water quality management, and they are open and available to public use and appreciation. The lands are being managed by the Monona Park District. And although the district currently does not actively manage the marsh complex, the administrative and managerial mechanisms exist for correcting any problems that might arise.

CONCLUSIONS

According to scientists of the Corps, the WDNR, and the WDOT, among others, the mitigation project was successful. Wetlands were restored to areas that had been filled or drained, and acceptable plant communities were established. The worst fears of the opponents to the highway project were not realized. Wetlands were created and the Yahara River Marshes and the associated plant and animal life survived and were even enhanced.

Figure 7-10 Restoration Area 4—high and shallow marsh and open water. Photograph by Elizabeth Day, Wisconsin Department of Transportation. Photo also appears in color insert.

The wetlands adjoining the project, but outside of the impact area, were not adversely affected by the highway. The restored and enhanced wetlands are considered successful, despite the fact that the plant communities that eventually developed in these areas were not exactly what the designers had intended. Still, given that this was the first wetland mitigation project undertaken by the WDOT, it represents an advancement in restoration science.

From one perspective, only the restored and created wetlands might be counted as a contribution to the wetland resources of the area. In this case, the gain to loss ratio would be 1:1, assuming that the gained wetlands function as well as those lost. However, wetland resources were also improved through enhancement and protected from future development, contributing to the region's resource.

Besides the physical contribution to the resource base, more was gained from the mitigation experience. This was the first time in Wisconsin that mitigation for wetland losses was required. It was the first time that the WDOT engaged in the protection and construction of new or enhanced wetlands. The construction techniques used in this effort were new and expensive. But despite the lack of experience, the mitigation was accomplished and the participants were satisfied that value was added to the regional environment.

The scientists and regulators involved in the mitigation process characterized the results as being successful. Both the FWS and the WDNR were satisfied with the protection and gain in wildlife habitat. This, in the end, was the goal and the primary wetland function considered both in the mitigation design and in the perceived value of the work. Flood control and water quality benefits are present, as demonstrated by the monitoring program.

Opposition to the project seems to have dissipated. The remaining opposition still questions the value of the South Beltline and laments its effects on urban sprawl. But of those interviewed, none characterized the mitigation project as being a failure. In fact, little formal public attention has been given to the mitigation wetlands since their construction. In the end, the Public Intervenor and the general public itself seemed to lose interest in the mitigation project, leaving the effort of restoring a successful landscape to the engineers and scientists of the state.

ACKNOWLEDGMENTS

A number of people had a hand in protecting the Yahara Marsh and advancing the idea of its importance to the environment. They helped to minimize wetland losses and to ensure that what losses occurred were properly mitigated. Early on, James Zimmerman, a professor with the University of Wisconsin, and his wife, Libby, along with a number of graduate students, most notably Barbara Bedford, highlighted the value of the Yahara Marsh and recommended its preservation. Zimmerman and his wife started the Wisconsin Wetlands Association and worked to draw public attention to the problems posed by a road traversing the marsh. In these efforts, they were joined by many others.

John Jackson, a biologist with the WDOT labored for a number of years to improve the design of the highway in order to minimize wetland losses. He, along with a team of engineers and scientists, negotiated the final mitigation agreement. Betsey Day, a graduate student at the University of Wisconsin and a biologist with the WDOT, prepared the mitigation design and helped shepherd the restoration work. She undertook construction supervision and base-line surveys. She also helped to evaluate the effectiveness of the various restoration techniques employed. Hal Meier, John Jackson's counterpart with the WDNR, initially opposed the project, but after it gained public support he worked closely with the other environmental scientists to ensure the best possible outcome. He worked many years on the resolution of the transportation issues and maintained a keen interest in the restored landscape. Katherine Falk, the Public Intervenor, helped to crystallize the issues over whether or not to build the highway and, in the end, her arguments appeared to set the stage for the final resolution.

John Jackson, Betsey Day, and David Siebert graciously helped the authors organize the materials and explain the facets of the Yahara River Marsh restoration. Their help is greatly appreciated.

REFERENCES

Ashworth, S. M., *Vegetation Analysis of the South Beltline Highway Wetland Restorations*, University of Wisconsin, Madison, WI, 1992.

Bedford, B. L., E. H. Zimmerman, and J. H. Zimmerman, *The Wetlands of Dane County, Wisconsin*, Dane County Regional Planning Commission, Madison, WI, 1978.

Boswell, T., *Wisconsin Wetlands Association: Born in a Freeway Fight*, Wisconsin Wetlands Association, Madison, WI, undated.

Crabtree, A., E. A. Day, A. Garlo, and G. Stevens, *Evaluation of Wetland Mitigation Measures*, Federal Highway Administration, McLean, VA, 1990.

Curtis, J. P., *The Vegetation of Wisconsin: An Ordination of Plant Communities*, University of Wisconsin Press, Madison, WI, 1959.

Day, E. A., *Mitigation for the South Madison Beltline: Dane County Wisconsin*, Wisconsin Department of Transportation, Madison, WI, 1986.

Durrie, D. S., *A History of Madison, the Capital of Wisconsin*, Madison, WI, 1874.

Falk, K., *Letter in Final Environment Impact Statement*, FHWA-WISC-EIS-83-01-F, Washington, DC, 1984.

Federal Highway Administration and State of Wisconsin Department of Transportation, *Draft Environmental Impact Statement*, Washington, DC, 1983.

Federal Highway Administration and State of Wisconsin Department of Transportation, *Final Environmental Impact Statement*, FHWA-WISC-EIS-83-01-F, Washington, DC, 1984.

Jackson, J. O., *Madison South Beltline Wetland Mitigation Project: An Evaluation of Wetland Planting and Seeding*, Wisconsin Department of Transportation, Madison, WI, 1990.

Owen, C., R. Quentin, J. Carpenter, and C. B. Dewitt, *Comparative Hydrology, Stratigraphy, Microtopography and Vegetation of Natural and Restored Wetlands at Two Wisconsin Mitigation Sites*, Institute for Environmental Studies, University of Wisconsin, Madison, WI, 1989.

Roherty, M., *Letter to District Engineer*, Army Corps of Engineers, St. Paul, Minnesota, Wisconsin Wetlands Association, 1985.

U.S. Geological Survey, *Water Resources Data—Wisconsin, Water Year 1996*, Madison, WI, 1997.

Wisconsin Department of Transportation, *Joint State/Federal Application for Water Regulators Permits and Approvals*, Army Corps of Engineers, St. Paul, MN, 1984.

CHAPTER 8

HOOSIER CREEK WETLAND

Driving southwest out of Denver, through the rapidly urbanizing Jefferson County, U.S. Highway 285 begins to twist and turn as it climbs toward Kenosha Pass. Urbanization is left behind as the road enters Park County, approximately 50 miles southwest of Denver. The highway runs between the steep escarpment of the Front Range of the Rocky Mountains and the North Fork of the South Platte River. Last improved in 1940, only two lanes existed along this stretch of road—each lane was 11 feet wide. The shoulders were narrow and unpaved and the tight curves made driving hazardous in both summer and winter. In 1986, improvements were proposed for a 2.9 mile stretch of road between Grant and Webster, Colorado (Fig. 8-1).

The need for the improvements was largely driven by safety—reducing sharp curves and widening and paving the shoulders—and by consideration of future demand. Despite the burgeoning population of neighboring Jefferson County, the rural population of Park County was not growing rapidly. However, the demand for access to recreational areas in Park County was increasing because of the population growth in Jefferson County and Denver. U.S. 285 serves as a major traffic artery for the metropolitan area.

The Colorado Department of Transportation undertook design of road improvements. Within the immediate planning horizon, additional traffic on the road was not expected, but because the road was being improved for reasons of safety, the Department concluded that widening the roadbed and adding 10-foot shoulders on both sides should be done as well, to avoid future environmental impacts and cost increases. To widen the roadbed, either the face of the escarpment on the north side of the road would have to be excavated or the North Fork of the South Platte River would have to be moved south. To avoid

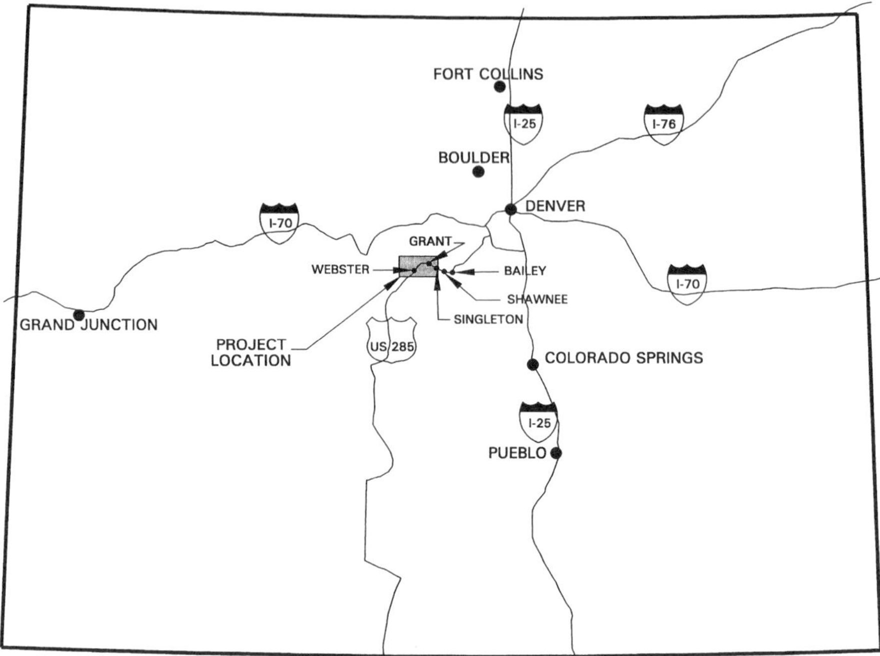

Figure 8-1 Project location.

possible rock falls and maintenance costs associated with excavation of the escarpment, the Department concluded that relocating the river would be less costly and, ultimately, less environmentally damaging.

Because the river flows through a narrow valley between the roadbed and an abandoned railroad embankment, very little space was available. To maintain the hydraulic capacity of the river, the channel section needed to be deepened and widened. This, in turn, caused the loss of 2.4 acres of wetland (Meiring et al., 1991).

The Department of Transportation proposed to mitigate the wetland losses both on and off site. Although the Department committed only to replace the wetland losses on a 1:1 basis, a greater area was restored or created. Approximately 1.5 acres of wetlands were created along the river corridor and an additional 3.21 acres were restored on Hoosier Creek, a tributary of the North Fork (Cooper, 1988). This latter, off-site mitigation is located approximately 1.5 miles north of the highway right-of-way but within the same watershed as the impacts. The actual mitigation ratio, therefore, was 1.5:1.

The design of the mitigation program was reviewed by the U.S. EPA, USFWS, the Corps, and the U.S. Forest Service (the landowner of the off-site mitigation property). After an extensive search of possible off-site mitigation locations, the Forest Service property was selected. It afforded a number of environmental benefits, some of which could not have been accomplished within the US 285 right-of-way.

The proposed improvements of the highway, the environmental impacts, and the proposed restoration were reviewed by public and private agencies and local citizens' groups as well. On the whole, there was little or no opposition to the improvements, although major concerns were raised about the temporary contamination of public water supplies downstream of the project area during construction. The Bailey Water and Sanitation District, located at Bailey, Colorado, pointed out to the Commander of the Omaha District of the Corps (the Omaha District has responsibility for the South Platte Watershed), that their water intakes were downstream of the construction activities and that they anticipated that ". . . such major construction would increase our turbidity problems immensely" (Painter, 1988). The Upper South Platte Water Conservation District recorded a similar complaint (Butler, 1988), but went on to propose that through the wetland mitigation work associated with the project, water quality conditions on the North Fork of the South Platte River could be improved: ". . . channel realignment and wetland construction associated with the Highway 285 improvement project can be modified to provide not only wildlife mitigation but also water quality mitigation provided all parties are willing to cooperate." The water quality problems alluded to were related to heavy metals—copper, lead, silver, and cadmium—being washed off nearby mining sites. The District proposed that the Department work with them to develop wetlands along the North Fork that could be used to remove or control the heavy metals. More specifically, the Shawnee Water Commissioner's Association, in Shawnee, Colorado, raised concerns about the disturbance of mine tailings within the project and the possibility that these tailings might be conveyed to its intake and into its water supply system (Tolpo, 1988).

Only the Park County Planning Commission raised concerns about the loss of wetlands (Bowman, 1988). They also joined the Upper South Platte Water Conservation District in suggesting that all agencies cooperate to expand the objectives of the wetland mitigation program to improve water quality along the North Fork. The planning commission also pointed out that Highway 285, through the Platte Canyon, "serves as a major route for metropolitan residents to reach and enjoy some of Colorado's finest outdoor recreation. Through this project, the possibility exists for a coordinated effort to begin between Colorado's Division of Highways and Division of Wildlife to improve the habitat within this corridor, possibly as part of the Governor's proposed 'Watchable Wildlife' program."

By February of 1987, when the Colorado Department of Transportation submitted the Section 404 permit application to the Omaha District of the Corps it had effectively worked with the regulatory and oversight agencies and reached an agreement on wetland mitigation. Consequently, issues raised as a result of the public notice were of minor significance and typically became recommendations for conditions on the permit. For example, the Division of Wildlife in the Colorado Department of Natural Resources noted that "after extensive coordination on this project with the Colorado Highway Department, we are satisfied that this proposal will have minimum possible negative impacts on wildlife/fisheries habitat. We have no objection to the issuance of this permit'' (Weber, 1988).

The FWS also endorsed the project, but asked that several conditions be included in the permit (Opdycke, 1988):

> Should mitigation of the 2.0 acres of wetlands [this was all that was needed although 3.21 acres was restored] proposed for Forest Service land along Hoosier Creek not occur for any reason, the Department should be required to find an alternate mitigation site as close as possible to the area of impacts and accomplish wetland mitigation in the amount originally proposed.
>
> Additionally, we recommend the permit be conditioned to require the Department to provide draft copies of all mitigation plans to the resource agencies for review and comment. Plans for the 0.5 acres to be mitigated on site, the 2.0 acres on Hoosier Creek, and the plans prepared by the Department's consultants for the remainder of the construction corridor should be made available to the concerned agencies.

Further, the FWS concurred with the Corps that there would be no effect on threatened and endangered species.

The Department of Transportation responded to all issues raised by local, state, and federal agencies and private citizens. The issues of increased turbidity and the transport of sediment-related solids were addressed directly and promises made to control the entrainment and transport of sediments. Erosion control plans were included in the construction documents as required by the 404 permit. The Department of Transportation agreed to the recommendations of the FWS, and the recommendations were incorporated into the permit. The Department, however, declined to become involved in a wider effort to create additional wetlands along the corridor to ameliorate the water quality problems and other issues presented by the various water supply districts downstream from the project site (McOllough, 1988). The Department justified their position with the following statement: "Federal Highway Administration policy does not allow for expenditures of funds for mitigation beyond what is required for impacts associated with our project.''

Design work by the Department began in 1986. By the end of 1987, the design and environmental issues had been resolved. In March of 1988, the Corps issued the public notice for the project. Although several requests were made, the Corps declined to hold a public hearing. Instead, the issues raised by the public and affected agencies were directly addressed in writing by the Department of Transportation. On June 10, 1988, the District Commander of the Corps issued permit #CO 2SB OXT 2 010629. Construction work was initiated in this same year, as was mitigation. After a field inspection on February 7, 1990 by the project manager for the Corps, the mitigation for the highway project was determined to be successful and complete.

ENVIRONMENTAL AND SOCIAL SETTINGS

The highway improvement project is located in north-central Park County, approximately 50 miles southwest of Denver. It is nestled in the valley of the North Fork of the South Platte River. The stream drains the western slope of the Front Range of the Rocky Mountains. The elevation range of the county is from 6,000 to nearly 12,000 feet. Grant, Colorado, the town situated within the project area, is at an elevation of 8,670 feet.

The project and the mitigation site are both in the Front Range. These mountains are immediately west of Denver, stretching from Rocky Mountain National Park on the north to Pike's Peak on the south. The western boundary follows the Continental Divide from Long's Peak to James's Peak (Weber, 1976).

Within this range a number of plant communities have been identified, classed physiographically by elevation, starting with the plains at approximately 5,700 feet above mean sea level (fmsl), moving through mesa and foothills up to 8,200 fmsl, to alpine tundra above 13,000 fmsl. The project area is situated within the subalpine class, between 8,800 and 11,000 fmsl. According to Weber (1976), this class is characterized by "Engelmann spruce, sub-alpine fir, and limber pine forest, interspersed with moist meadows, ponds, and bogs. Very rich in wildflowers."

The mean temperature in Grant is 38.7°F, the mean daily maximum temperature is 53.3°F, and the mean daily minimum temperature is 24.1°F. January, the coldest month of the year, has a mean temperature of 20.9°F. July, the warmest month, has a mean temperature of 58.8°F. With a 70% probability, the growing season for agricultural crops is approximately 83 days, extending from June 16 through September 6. As Cooper (1998) has noted, however, the growing season for native, subalpine plants, including hydrophytes, is a good deal longer because these plants have adapted to colder air and soil temperatures.

Annual precipitation in Grant is 15.5 inches with a 30% probability that the annual total might be <11.5 inches or >17.4 inches. Approximately 35% of the total amount of precipitation, or 5.4 inches, occurs as snowfall. The wettest month of the year is July, at 2.49 inches of precipitation. This month is closely followed by August, at 2.43 inches. The driest month of the year is January, at 0.45 inches, with February being a close second at 0.53 inches. In the highest stream discharge period, May through August, 52% of the precipitation occurs.

The watershed of the North Fork at Grant is approximately 127 square miles. The northwestern boundary of this drainage area falls along the border between Summit and Park Counties. After adjusting for the discharge into the North Fork from the Harold D. Roberts tunnel, which supplies water to the Denver Metropolitan area, the mean annual flow is 71.6 cubic feet per second (cfs) as measured below Geneva Creek at Grant, Colorado (U.S. Geological Survey, 1996). The annual yield is 7.7 inches. Streamflow accounts for approximately 49% of the precipitation that falls on the watershed. Streamflow is poorly distributed through the year, somewhat reflecting precipitation patterns. More than

65% of the streamflow occurs over the 120 days, May through August, of the snowmelt season. The other 35% occurs in the remaining months, September through April. This poor distribution is due in large part to snow accumulation at the higher altitudes and snowmelt during the late spring and early summer. The highest monthly mean flow, 294 cfs, occurs in June, and the lowest mean monthly flow, 38.6 cfs, occurs in March.

The high percentage of runoff relative to precipitation occurs primarily because of the low evapotranspiration due to moderate to low temperatures and the very steep slopes of the mountainous watersheds. Also, the inorganic soils on the mountain slopes have very low water retention capability. In the alpine valleys, on the other hand, organic matter tends to accumulate and the peaty soils have a greater capacity to retain water.

The general soil association for the area is Herbman–Hiwan: "moderately sloping to very steep, shallow, well-drained, stony, gravelly, loamy, and sandy soils that formed in the material derived from igneous and metamorphic rocks" (Price and Amen, 1984). These soils are found on the side slopes and ridges of mountains, slopes ranging from 5% to 70%. Coniferous trees and shrubs and grasses account for most of the vegetation, and the soils exist mainly in the elevation range of 7,600 to 10,000 feet. This soil association covers approximately one-third of the surface area in Park County. Of this, 30% are Herbman soils, 20% Hiwan soils, and the remaining 50% are composed of rock outcrop and other minor soil units. The Herbman soils are found on mountain slopes and ridges, are shallow and well drained, and are derived mainly from igneous and metamorphic rocks. Hiwan soils are also found on mountain slopes and ridges, mainly facing north. They are shallow and well drained and are underlain by very gravelly sand. Weathered hard rock is at a depth of 5 to 20 inches. The minor soil units associated with Herbman–Hiwan are rock outcrops on ridges and backslopes, Grimstone and Peiler soils on north-facing mountain slopes, Kitteredge and Troutdale soils on mountain slopes and in drainageways, and venerable soils on low terraces. These soils mainly support forestry, grazing, wildlife habitat, and recreation. They also could be used for community development. Within the mountain valleys, the soil association underlying most wetlands is Cryoborolls–Cryaquolls. This soil association contains sandy–gravelly loam covered by a dense mat of partly decomposed vegetation. It is composed of poorly to well drained soils with a moderate capacity to hold water (Price and Amen, 1984), which is advantageous for the drier months of the year.

Within the Rocky Mountain region, four major wetland types have been identified (Cooper, 1989):

> Riparian (riverine) Wetlands—Found along moving water courses such as rivers and creeks, these wetlands receive a large seasonal pulse of water from the melting of mountain snowpacks. Flooding, sediment erosion, and deposition are characteristic. Riparian wetlands can be forested, such as the well-known Cottonwood Gallery Forest in the lowlands; shrub dominated, such as the willow thickets

found along many streams; or dominated by herbaceous flowering plants, such as those along cascades in the mountains. Many riparian wetlands have saturated soils and/or high water tables only early in the growing season.

High Mountain Wetlands—These occur in regions that were glaciated during the Pleistocene Period. The glaciers have carved the mountains and deposited till (rocky material pushed ahead or to the side of glaciers, or left when a glacier melts out), creating landforms that slow the runoff of water. Wetlands are abundant and may occur behind glacial terminal moraines, where a valley is flat and streams meander, in kettle ponds within moraine deposits, and where glaciers have impounded streams. Many high mountain wetlands have peat soils that are saturated for most of the growing season.

Basin Wetlands—Occurring in the level intermountain regions, these may be closed basins such as Great Salt Lake in Utah, where surface runoff from the mountains collects and no drainage occurs, or they may be smaller wetlands, such as those in the intermountain parks and basins of Colorado, Montana, and Wyoming. Many basin wetlands are saline or alkaline because of the chemical characteristics of the solids transported into the basin by surface water. When the water evaporates, the saline or alkaline solids remain.

Urban Wetlands—In urban, commercial, and industrial areas, these wetlands are either created by runoff from hard surfaces or are fragments of naturally occurring wetlands that have been influenced by the heavy loads of nutrients, pesticides, herbicides, metals, petroleum products, and other urban pollutants.

The class of Riverine Wetlands applies to the wetlands affected by the project and to the mitigation sites.

In general, the wetlands of Park County, including the project and the mitigation sites, have been affected by mining, industrial, and agricultural activities. Owing to mining activities, sediment loads were conveyed downstream and settled out in the slower-moving waters in the valley meadows. These sediments often altered the form and function of the impacted wetlands. In other cases, waste products resulting from the production of commercial products such as charcoal, and ore processing were deposited in wetlands. Agricultural activities, such as grazing, contributed to surface and stream bank erosion. Cattle removed vegetation, exposing the soil to wind and water erosion. Trapping reduced the beaver population, curtailing the natural stabilization of wetland complexes and stream channels. With the loss of beaver and cattle traversing the stream channel, downcutting occurred. These activities all led to the direct loss of upstream wetlands and sedimentation of downstream wetlands.

Although there are no statistics on wetland losses within Park County and the immediate vicinity of the project, Colorado ranks 21 out of 50 states in terms of the percent of wetlands lost. From 1780 to 1980, wetland losses in the state were estimated to be approximately 50% (Dahl, 1990). Wetland losses in Park County are likely far less than the statewide average because of the large undeveloped land area in the county. However, wetlands have been lost to water development (storage reservoirs) and mining activities, including peat mining (Cooper, 1990).

The land use of Park County reflects the low population density. Seventy percent of the land is public, within national forest boundaries, 25% is used for ranching, and the remaining 4% to 5% is used for mining, industrial, and other uses. The public lands are used extensively for hiking, skiing, fishing, and other outdoor activities. Tourism is the principal industry of the county.

Park County, 2,166 square miles, was established in 1861. By 1900, the population had reached approximately 3,000, with a population density of 1 person per square mile. By 1920, the population had dropped below 2,000, rose to 3,000 by 1940, fell by 1960 to below 2,000 again, and since then has steadily risen. Today the population is 7,200, or approximately 3 persons per square mile (Fig. 8-2). In contrast, Jefferson County, immediately to the east of Park County and west of Denver, has urbanized substantially from the 1950s to the 1990s. This county, in 1900, had a population of around 9,000. By 1950, the population had grown to close to 56,000, or 71 persons per square mile. From 1950 to 1990, the population of this urbanizing county grew to 438,000, or approximately 558 persons per square mile (Fig. 8-2).

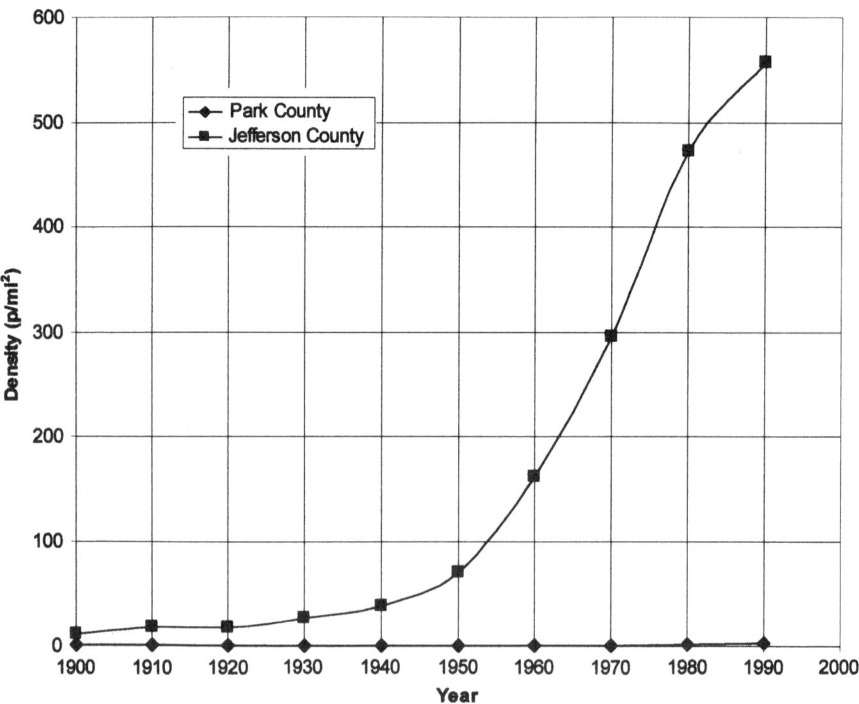

Figure 8-2 Population densities of Park and Jefferson Counties, Colorado.

THE PROJECT AND ITS WETLAND IMPACTS

The highway improvement project began immediately west of the small town of Grant, Colorado, and ended 2.9 miles southeast of town. Across this distance, the elevation of the roadway descends from 9,016 to 8,601 fmsl, resulting in an average grade of −2.9%. The road improvements were intended to meet the design standards, including traffic projections, for the subsequent 20 years. This involved adding a 10-foot wide shoulder to each side of the highway, constructing the roadbed for two future lanes, and improving five horizontal curves (Fig. 8-3a–d).

To expand the width of the road and improve the curves, the roadbed had to be widened. For most of this stretch of highway, the northern side of the road is positioned against the valley wall, with the southern side abutting the North Fork of the South Platte River. To avoid landslide hazards and associated safety issues, and after considering a number of alternatives, the Colorado Department of Transportation and the Federal Highway Administration concluded that it would be far safer to expand the roadway into the floodplain and streambed and move the stream further south (Fig. 8-4). To accomplish this task, approximately 4,100 feet of stream channel had to be relocated, filling 2.4 acres of wetlands. The wetlands were predominantly shrub carr (a wetland characterized by woody shrubs, organic (peat) soils, and an abundance of water), riverine wetlands. They were moderate in quality, having been disturbed over the years by upstream mining and forestry and by road and railroad construction. An abandoned railroad bed lies on the southern edge of the North Fork floodplain across the stream from U.S. 285. The affected wetlands were dominated by mountain willow (*Salix monticola, Salix ligulifolia*), river birch (*Betula fontinalis*), bluejoint reed grass (*Calamagrostis canadensis*), and various sedge species (*Carex* spp.).

Given the limited, available non-wetland areas along the stream, the Department found it possible to mitigate only 0.5 acres of wetlands within the corridor. Consequently, only one-half acre of in-kind wetland mitigation was planned. The mitigation wetlands were to be shrub carr, dominated by willow.

The remaining 1.9 acres were mitigated outside of the highway corridor. The regulatory agencies made it clear that the off-site mitigation must be as close as possible to the sites of impact. After searching for a suitable site for more than a year, the Forest Service came forward with a proposal to restore a wetland on their property. The location, within the Pike National Forest, was along Hoosier Creek, which is about 1 mile west of the start of the road improvements and approximately 0.5 mile north of U.S. 285, on Park County Road 58 (Fig. 8-5).

The Hoosier Creek site had been filled in the 1880s and 1890s, when the upland forests had been logged to produce charcoal for smelters in the nearby mining camps and towns, such as Drake, Colorado (Hand and Anjulshi, 1988). Over 28 "beehive" brick kilns were built and operated to meet the demand for charcoal. In the process, considerable waste by-products were produced.

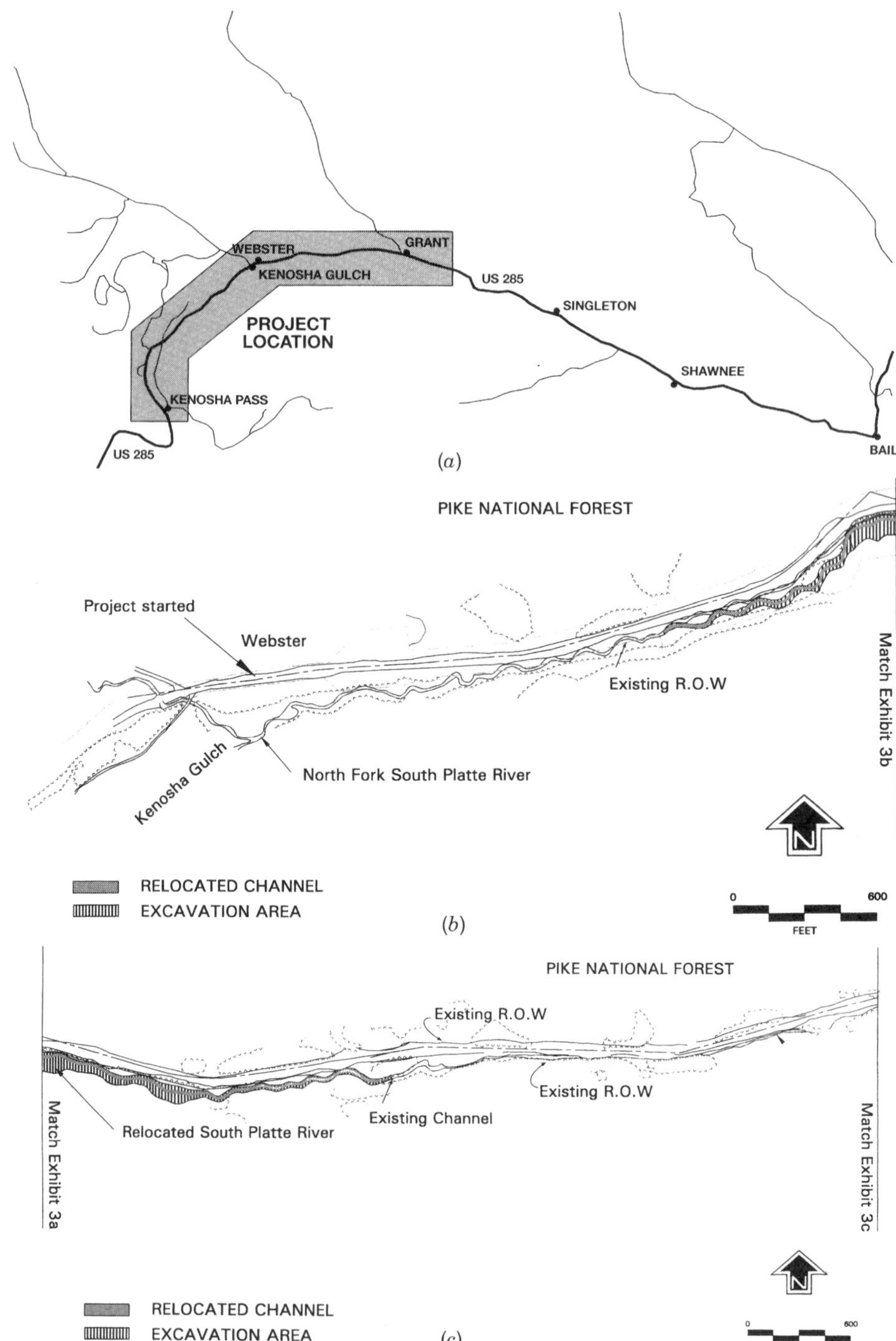

Figure 8-3 (*a*) Project corridor; (*b–d*) Alignments.

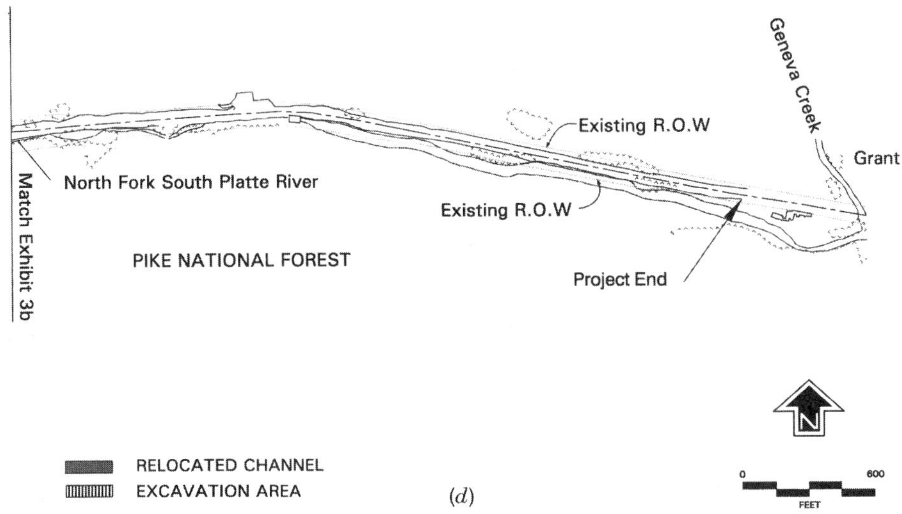

(*d*)

Figure 8-3 (*continued*)

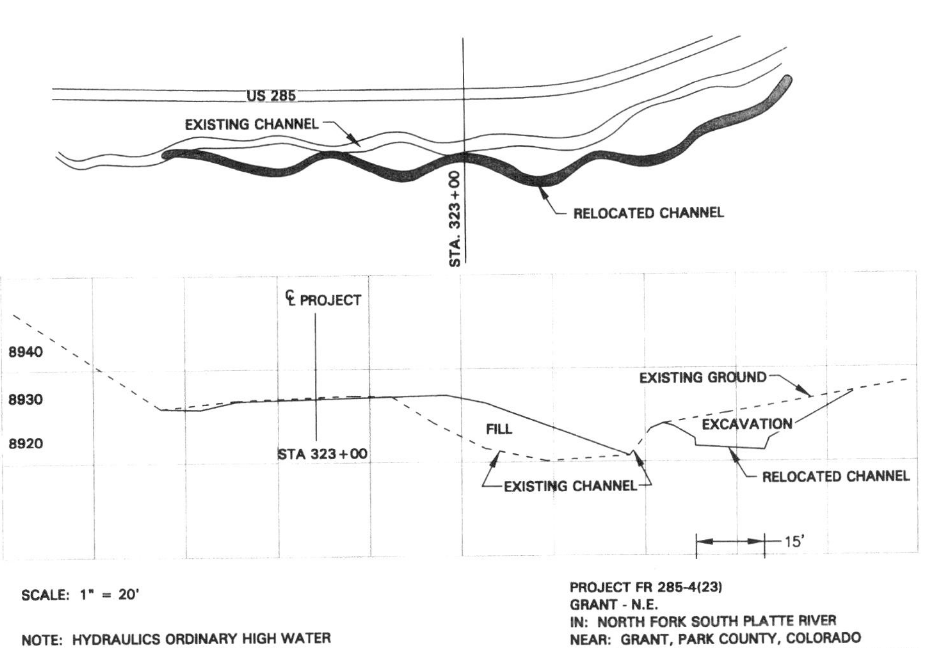

Figure 8-4 Road cross-section showing relocation of river channel.

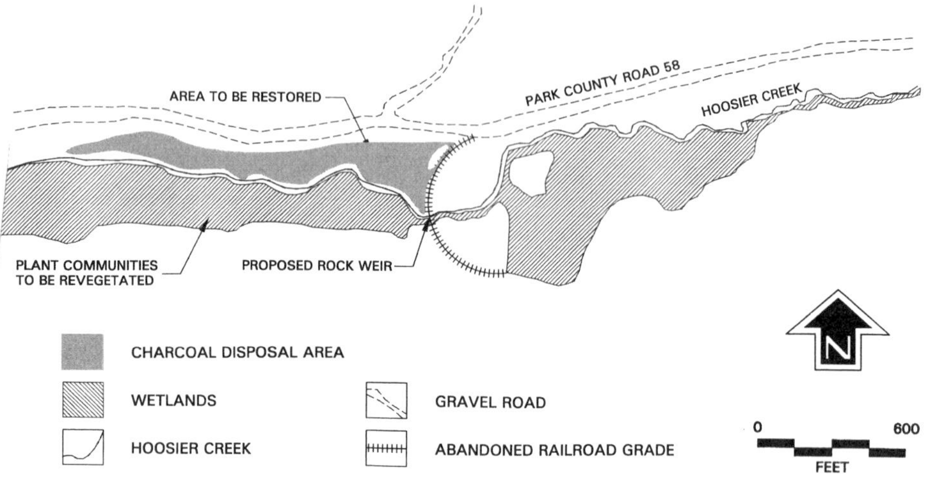

Figure 8-5 Hoosier mitigation site.

These wastes were dumped into the riparian wetlands on the north side of Hoosier Creek—a disposal technique used for approximately 10 to 20 years before the site was abandoned (Fig. 8-5). At the lower, southern end of the site, a railroad embankment was built and a trestle constructed over Hoosier Creek. The railroad was operated by the Denver, South Park, and Pacific Railroad and used to transport the finished charcoal. The railroad was ultimately abandoned in the late 1930s and the structure over Hoosier Creek removed.

Subsequent to the charcoal operation, the site was heavily grazed by cattle permitted on the site by the U.S. Forest Service, contributing to the erosion of the stream channel. On the south side of Hoosier Creek, the floodplain was not filled or heavily damaged by grazing. Wetland plants survived on the peaty soils. With the expectation that the surviving plant communities could be recreated on the north side of Hoosier Creek, the U.S. Forest Service proposed to the Colorado Department of Transportation that the site be used for mitigation. After careful consideration by and discussion with the regulatory agencies, the department proposed, in its Section 404 permit application to the Corps, that the two acres of off-site mitigation be accomplished at this location.

The mitigation plan called for construction of a rock weir (Fig. 8-6) at the abandoned railroad crossing at the east end of the site, raising the elevation of the water by 8 to 10 feet. The eroded channel above the control structure would be filled with excess angular rock obtained from the highway construction. The 10,000 cubic yards of charcoal waste along the north side of the creek was to be excavated and removed from the site. The material was transported to a nearby Boy Scout camp and used as the base for a parking lot. The charcoal was to be removed down to the peat, which exploratory borings had shown

Figure 8-6 Constructed rock weir in abandoned railroad embankment. Photograph by Kristine Meiring. Photo also appears in color insert.

was still present. Willows were intended to be the dominant plant species used in the restoration. They were to be the anchor of the shrub carr community. The willow plantings were to be established by taking cuttings of nearby mature willows and inserting them 12 to 18 inches into the peat and adjacent soil.

The plan was not specific about the plant species, other than willow, that ultimately might be in the restored wetland. Nor was it specific about the density or diversity of plants. There were no goals or objectives or specific criteria established for the restoration work. Further, no construction drawings or specifications were prepared, nor were they required for the permit application. But these items were prepared for the on-site mitigation because it was a part of the construction activity associated with the road improvements—plan drawings showed the location of the new stream bed and the placement of rock materials to control pools and riffles as well as the placement of vegetation along the new stream corridor. In regard to the off-site mitigation, no hydrologic studies, either surface or groundwater, were conducted prior to developing the plan. As a result of public comments on the 404 permit, hydrologic data were collected along the stream corridor to support future mitigation planning for the larger U.S. 285 corridor. Only after the beginning of the restoration effort were groundwater wells installed at the Hoosier Creek site and plant lists developed.

MITIGATION IMPLEMENTATION AND RESULTS

Construction of the new stream corridor and on-site mitigation along the North Fork of the South Platte River proceeded apace with the construction of the road. The contractor was overseen by inspectors from the Colorado Department of Transportation. The existing plan drawings served as a guide. The wetland soils were removed from the areas to be covered by the wider roadway and from the portion of the stream which was to be relocated. These materials were stockpiled for a short duration and then respread in the floodplain of the reconstructed channel to create wetlands. Also, some materials such as boulders, cobbles, and topsoil that were not used for the on-site mitigation were stockpiled for use in the off-site area. Existing vegetation, including trees, was to be left along the new channel edge adjacent to the roadway to provide a buffer between the road and the new channel to minimize noise and visual impacts. Also, the buffer vegetation, including plants from the seed bank, was to serve as a filter for highway runoff. Finally, a ditch was created parallel to the highway to provide detention for sediment removal prior to road runoff reaching the stream.

In October 1988, a restoration plan was prepared by Cooper (1988) for the Hoosier Creek mitigation site. In preparation of the plan, the proposed site was surveyed for wetlands and other resources that might be affected by the restoration work. The existing wetlands were mapped (Fig. 8-5), but the Corps did not require a 404 permit for the off-site mitigation itself, even though some of the existing wetlands were affected by site grading and by the rock weir, which was built across the stream channel.

The watershed tributary to the Hoosier Creek restoration site is small but steep. It encompasses 2.9 square miles and falls from 10,500 to 9,700 fmsl. The average slope is 0.065 ft/ft. The restoration site is located near the outlet of the watershed just upstream of its confluence with the North Fork. Prior to restoration, the slope of the stream was moderate to steep through the restoration site, although beaver had constructed dams upstream of the eroded channel section.

The south floodplain supported a sedge plant community, and the north floodplain contained the charcoal waste. The thickness and extent of the charcoal ash was determined in order to assess the amount of excavation necessary to expose the surface of the predisturbance peat soils. Five monitoring wells were installed to assess the groundwater across the buried soils. Based on the observed levels and the extent of peat, Cooper (1988) estimated the amount of wetland that could be restored. He went on to speculate that:

> The charcoal ash in the Hoosier Creek Valley was placed onto willow carr wetlands. These wetlands most likely supported beavers and were quite complex in their hydrology and vegetation, just as the willow carr wetlands are in the upper part of Hoosier Creek Valley. The soil profile exposed by the degradation of Hoosier Creek into its floodplain shows the presence of peat deposits 125–150

> cm (4–5 feet) thick beneath the charcoal ash. In the southern Rocky Mountains, peat this thick indicates a very long period of wetland stability, most likely many thousands of years.

In the end, Cooper estimated that there had originally been 6.75 acres of wetland downstream of the railroad embankment (Fig. 8-5) and 6.3 acres upstream. Those former wetlands covered by charcoal comprised 3.2 acres. Besides the presence of peat, the groundwater regime indicated the appropriate hydrology for peat wetlands, those that once existed at the site. The groundwater was within 145 cm of the surface in July and 149 cm in September.

Two critical recommendations were set out in the restoration plan of Hoosier Creek: remove the charcoal ash from the buried wetland surface, and construct a hydraulic control structure across Hoosier Creek above the old railroad grade. The plan (Cooper, 1988) provided additional recommendations as follows:

- The excavation work should be done in the late fall or early winter of 1988. Care should be taken not to run heavy equipment over the exposed peat surfaces. (This work was done in the winter when the peat was frozen and equipment could operate on the surface.)
- Care should be taken to expose only the top peat surface and not remove any of the peat. The goal was to expose the soil surface that was buried by the charcoal.
- Excavation should begin at the edge of Hoosier Creek and work back toward the upland edge. This would minimize the chance that heavy equipment might compact the peat surface.
- Hydraulic control structures should be built in the area west of the railroad grade to stop the degradation of Hoosier Creek and bring the water table in the peat areas adjacent to Hoosier Creek back up to the ground surface elevation.
- Consideration should be given to building one drop structure for each 12 to 18 inches of relief upstream from the grade.
- The first drop structure should be built just upstream from the railroad grade and should have a top elevation a few inches higher than the surrounding banks in the center of the creek. The structures should be buried in the banks.
- The first drop structure should be the most durable and probably should be built of logs and rocks, but the construction specifications should be determined by the Department of Transportation. Three or more structures may be necessary to stabilize the creek and bring the water table up to near the ground surface. The goal is to develop a valley system with a high water table. The water table does not have to be at or above the ground surface over the entire restoration area, but it should be close to the ground surface for the early portion of the growing season in all areas.

The water table can drop to 1 to 3 feet below the soil surface in the later portion of the summers in many portions of the restoration area.

- The peat surface should be planted with willow (*Salix* spp.) stem cuttings. These should be from dormant, living willows located in Hoosier Creek. Cuttings should be from different shrubs in different parts of the valley so that many different willow species will be introduced into the wetland restoration site. Cuttings should be taken while the willows are dormant, but when the ground in the mitigation site is not frozen.
- Cuttings at least 18 inches in length should be cut at an acute angle, with sharp pruning shears, from the top of willow branches. These cuttings should be placed in a bucket of water to keep them moist during the cutting operation.
- Willows should be planted in a density of approximately one stem per 10 square feet. The willows should not be planted in an evenly spaced pattern to resemble a grid, but should be planted in an irregular pattern to make a more complex and interesting site.
- No other planting or seeding should be done.

The plan was quite simple and yet detailed in its implementation. After further site evaluation by the engineers of the Colorado Department of Transportation, only one hydraulic structure was constructed—at the railroad grade. It was built from stone and the original design was slightly modified a year later after icing problems developed on the downstream face of the structure. The structure changed the grade of the creek through the site from 0.06 ft/ft to 0.02 ft/ft. The charcoal wastes, which were required to be excavated, were excavated and hauled a short distance up the northern valley slope. Ultimately, the wastes were capped with local soils to prevent their movement down the slope.

The exclusive use of willows was proposed because of the difficulty of finding seeds or plant materials for other species found in subalpine wetland systems. Seed sources were not readily available when this project was undertaken. Further, it was believed that the other seeds would find their way into the restoration site by means of wind, water, wildlife, and rhizome migration from the extant, native wetland adjacent to the restored area. Also, the seed bed in the exposed peat was expected to germinate. The plan also anticipated some involvement of beavers: "Some beaver activity is natural to expect, but the removal of a large percentage of the planted willows could slow down the establishment of the wetland. If this occurs, it might be necessary to live trap one or more beavers from the area and transplant them to other areas" (Cooper, 1988).

The Department of Transportation's site engineer, Ed Demming, took a personal interest in the restoration effort. Considering Cooper's recommendations and the stream's hydraulics, he carefully organized the construction procedures for the control structure at the railroad grade. He personally supervised the selection of materials and placement of stones to emulate a naturally occurring steep riffle. When construction work began in January of 1988, he oversaw the backfilling of the eroded channel and the grading of the landscape following Cooper's plan.

After the spring thaw and runoff in May 1988, the willow cuttings were planted. Approximately 10,000 cuttings of several willow species were prepared, treated with Rootone (a commercial product intended to promote root development), and inserted into the newly exposed peat soil. Construction and planting were completed in the fall of 1989.

In the following year, the results of the first growing season were evaluated. Photographs and video recordings were used to assess visually the early and late season conditions. Ten groundwater wells were established and water levels measured when vegetation surveys were conducted. In addition, eighteen 2- by 4-meter quadrats were established (Fig. 8-7) based on the following conditions:

1. Willow cutting enumeration
   a. Number of cuttings producing shoots
   b. Number of cuttings producing roots
   c. Number of shoots per stem
   d. Average shoot length
   e. Longest shoot length

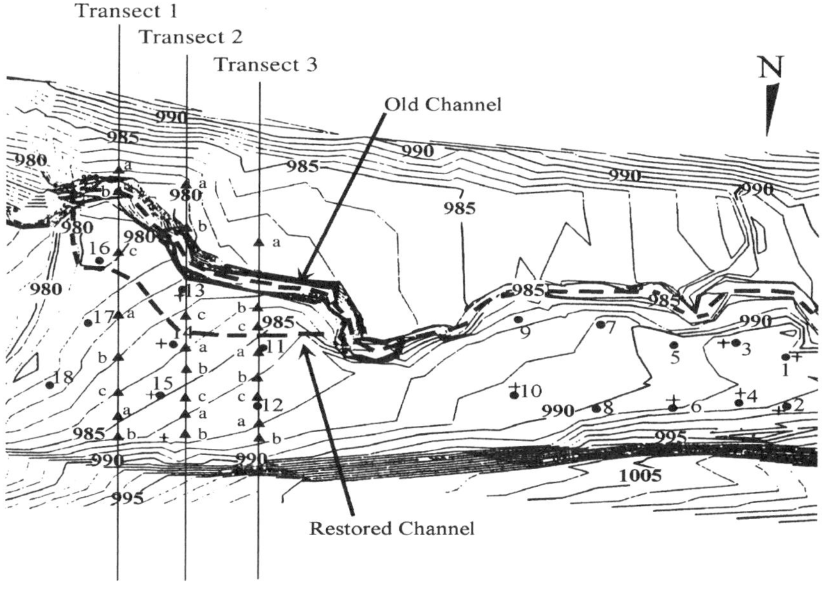

Figure 8-7 Plant sampling plots (1990) and transects (1991) and groundwater monitoring wells. (From Von Loh, 1991.)

2. Herbaceous vegetation
   a. Source (sod or seed)
   b. Ocular estimate of cover

The results of the first growing season were assessed (Meiring et al., 1991):

- A majority of the willow cuttings planted in the 15 study plots produced both roots and shoots (98% and 86%, respectively) by July, 1989.
- The number of shoots per stem averaged from two on drier sites to 16 on wetter sites.
- The longest shoots approached 18 cm in length, averaging approximately 10 cm long.

In July, herbaceous species covered only a small portion of the study plot area. The annual willow herb (*Epilobium paniculatum*) was the most successful colonizing herbaceous species observed across the site.

Similar results to those recorded in July were recorded in August, 1989:

- Root and shoot development occurred on 95% and 90%, respectively, of the willow cuttings which were monitored.
- The number of shoots per stem remained at five for the cuttings monitored.
- The average shoot length more than doubled over this 1½ month period. The length of the longest shoot per cutting on average nearly tripled over this time period.
- The longest shoot length recorded on the study plot was 61 cm.
- Herbaceous species generally constituted less than 5% of the total cover in the study plots.

Species of willow herb and tufted hair grass (*Deschampsia caespitosa*) and sedges (*Carex* spp.) were the most common invaders of the early exposed surface during the first growing season.

By July of 1989, beaver had begun to construct dams through the restored wetland and along the new drainage course. These dams were expanded by August. One relatively large pool and two smaller ones were evident on the site. Also, signs of elk were observed in the restoration area despite a perimeter fence, which was intended to keep cattle out. The elk had browsed on the new willow shoots and had pulled a few cuttings completely out of the peat substrate.

After the 1989 growing season and based on the available information, the Corps approved the mitigation as it existed. The Corps' official record noted that the complete scope of the work was in compliance with the permit, that all the standard and special conditions had been met, that planting had been completed at the Hoosier Creek site, and that relocation of the North Fork was complete. This approval by the Corps, however, did not stop the monitoring program on the Hoosier Creek restoration site. The Department of Transportation continued for an additional 2 years (1990 and 1991).

The 1990 monitoring report (Von Loh, 1990) made a number of observations germane to the long-term success of the newly formed landscape. In general, the site was affected by several factors, including the following:

- A prolonged drought at the beginning of the growing season in June of 1990.
- Subsequent abnormally wet conditions in July and August.
- Additional dam building activity by beaver.
- Browsing of willow shoots by elk.

The drought conditions during the onset of the growing season affected some willow cuttings along the upper, drier margins of the wetlands and adjacent mineral, upland soils. In the driest areas, which were considered marginal for willow establishment, a majority of the cuttings died. The depth of groundwater at these sites, measured in late August, was 1 to 18 inches. At slightly lower elevations on the site, where late August groundwater levels of 8 to 10 inches were recorded, the shoots on willow cuttings established during the 1989 growing season were dead, but the cuttings had resprouted from the base. Some of these new sprouts exceeded 15 cm in length, as observed informally during the monitoring visit. The mortality among willow cuttings was the result of inundation and persistently high groundwater. This mortality was evident within and adjacent to the large beaver pond, west of the beaver pond in the main channel area, and near the old railroad grade where pond water was present. Groundwater was either at or above the surface in these areas or within 2 inches of the surface in late August. Some willow cuttings had been incorporated into the beaver dam and resprouted there. Undoubtedly, a few cuttings were also used for food by the resident beavers.

As expected, the contribution of herbaceous species to the plant cover increased significantly over the second growing season. Although only a trace of herbaceous cover was noted in 1989, an average of approximately 30% cover was reported in the established plots in 1990. The range of cover values varied from approximately 10% over non-wetland, mineral soils to approximately 60% on wetland soils dominated by tufted hair grass (*Deschampsia caespitosa*). Species diversity increased with an average of 12 herbaceous species present per plot and a range from seven species occurring on mineral soils to as many as 20 species in the plot containing the exposed peat (Table 8-1).

Wind-blown seed appears to be the primary mechanism for early reinvasion of these wetlands. Incidental pieces of sod left following the excavation of the charcoal residue were the next most important contributor of plant species. Undisturbed sedge communities along the south side of the excavation sent rhizomes as much as 1.5 m into the reclaimed area. In a similar manner, a stoloniferous cinquefoil (*Potentilla* spp.) was quite successful in colonizing the

TABLE 8-1 Relative Abundance of Herbaceous Species on the Restored Study Plots

Plant Species[a]	Growth Form	Relative Abundance[b]
Agropyron sp.	Grass	3
Alopecurus aequalis	Grass	2
Deschampsia caespitosa	Grass	5
Festuca arizonica	Grass	2
Lycurus phleoides	Grass	2
Poa sp.	Grass	4
Carex utriculata	Sedge	1
Carex sp.	Sedge	3
Juncus arcticus	Rush	4
Artemisia frigida	Subshrub	1
Pentaphylloides floribunda	Shrub	3
Achillea millefolium	Forb	2
Antennaria rosea	Forb	2
Aster sp.	Forb	1
Capsella bursa-pastoris	Forb	1
Caryophyllaceae	Forb	1
Chenopodium sp.	Forb	2
Cirsium sp.	Forb	2
Corydalis aurea	Forb	1
Cruciferae	Forb	1
Descurainia sophia	Forb	1
Draba sp.	Forb	2
Epilobium paniculatum	Forb	5
Epilobium sp.	Forb	5
Heterotheca sp.	Forb	1
Lomatogonium rotatum	Forb	1
Lepidium densiflorum	Forb	1
Mimulus sp.	Forb	1
Penstemon sp.	Forb	1
Polemonium sp.	Forb	1
Polygonum aviculare	Forb	5
Potentilla sp. (stolon)	Forb	4
Potentilla sp.	Forb	3
Prunella neglecta	Forb	1
Ranunculus cymbalaria	Aquatic	2
Ranunculus hyperboreus	Aquatic	4
Rumex sp.	Forb	2
Sedum lanceolatum	Succulent	1
Taraxacum officinale	Forb	2

[a]Note: Includes only the species that occur in the study plots.
[b]1 = Uncommon, 3 = Common, 5 = Abundant.
Source: Van Loh, 1990.

site (Meiring et al., 1990). All of the ponded areas supported buttercup (*Ranunculus hyperboreus*) in the more shallow waters. This species and *Ranunculus cymbalaria*, which grows on muddy areas adjacent to shore lines, formed a nearly solid mat.

After the second growing season, almost 60% of the willow cuttings occurring in the study plots survived (Von Loh, 1990). The majority of the cuttings that did not survive succumbed to inundation. Shoot lengths were recorded, but these data appear to have less importance because, as the shoots branched out, they were browsed. The average shoot length increased by approximately 50% site-wide over the previous season (14.4 cm versus 21.8 cm). The average length of the longest shoot remained about the same whereas on some individual plots, the average length of the longest shoot doubled during the second growing season. Although willows were surviving, the site was becoming more of a sedge meadow than shrub carr. Moreover, owing to the activities of beaver whose dams had imposed a pool and riffle hydrologic regime, a great deal more water stood on the site than had been anticipated.

Wildlife use increased. In addition to the elk and the beaver, who extended the length of their primary dam, the site also supported nesting mallards and a pair of blue-winged teal. An unidentified shorebird nested on site and was observed with five young. Other wider-ranging species, including mule deer and red-tailed hawk, foraged in and near the site. The ponded areas supported an abundance of macroinvertebrates (Von Loh, 1991).

For assessing plant community development in the third and final year, 1991, three transects, oriented north to south, traversed both the left and right bank floodplains and the drainage ways in between (Fig. 8-8). The data resulting from the surveys along these transects were used to augment the data of the preceding 2 years, although the collection strategy was different. Along each transect, vegetation cover was estimated by species within 1 × 2 meter plots on alternating sides of the transect line. In this year, cover estimates were made for bryophytes in areas of open water within plots on peat, whereas bare ground and litter estimates were made for plots on mineral soils. The average cover values for vegetation along the three transects are: transect 1 (hair grass)—88% reclaimed and 108% undisturbed; transect 2 (sedge)—103% reclaimed and 80% undisturbed; and transect 3 (willow)—89% reclaimed and 113% undisturbed. (Cover values exceeding 100% result from multiple layers.) The cover values determined in this growing season, for herbaceous species only, ranged between 64% and 110%, nearly double that observed in 1990 (Tables 8-2 to 8-4).

The more notable differences between the restored and existing plant communities along transect 1 included the dominance of shrubby cinquefoil and sedge in the existing communities. Seedling plants of shrubby cinquefoil were present in the restored portion, which will eventually contribute a greater amount of vegetation cover as they mature. Restored portions of transect 2 remained saturated through the growing season, largely due to the influence of the beaver dam. As a result, species of sedge, by means of rhizomes, were

TABLE 8-2 Percent Cover of Reclaimed Wetlands North of Restored Hoosier Creek[a]

	Transect 1				Transect 2				Transect 3		
Species	Quadrat[b]: a	b	c		Quadrat[b]: a	b	c		Quadrat[b]: a	b	
Carex aquatilis	—	—	—		15	45	55		—	10	
Carex rostrata	2	2	—		15	20	10		—	—	
Carex sp.	—	1	10		5	5	—		—	—	
Cirsium sp.	4	—	—		—	—	—		—	—	
Deschampsia caespitosa	8	20	35		15	15	10		15	10	
Epilobium sp.	15	1	2		10	15	5		15	8	
Hordeum brachyantherum	—	4	2		—	—	—		5	1	
Juncus arcticus	5	1	2		—	—	3		2	—	
Pentaphylloides floribunda	3	—	—		—	—	—		—	—	
Phleum alpinum	5	5	12		—	—	—		2	2	
Poa sp.	30	20	10		—	—	—		—	2	
Potentilla sp.	15	5	10		5	3	5		2	6	
Salix sp.	—	2	15		5	—	8		35	45	
Unidentified forbs and grasses	2	5	12		11	3	5		2	2	
Bryophytes	—	—	—	Avg.	15	3	2	Avg.	10	4	Avg.
Total Cover	89	66	110	= 88.3	96	109	103	= 102.7	88	90	= 89
Open Water	—	—	—		5	—	2		—	—	

[a]These wetlands are re-forming over peat saturated to the surface during the entire year along transects 2 and 3, but only during the spring along Transect 1.

[b]Quadrat size was 1 × 2 m.

Source: Von Loh, 1991.

TABLE 8-3 Percent Cover of Undisturbed Wetlands South of Restored Hoosier Creek[a]

Species	Transect 1 Quadrat[b]: a	b	c		Transect 2 Quadrat[b]: a	b	c		Transect 3 Quadrat[b]: a	b	c	
Achillea millefolium	—	—	—		2	—	—		—	—	—	
Betula glandulosa	—	—	—		—	—	10		—	—	—	
Calamagrostis inexpansa	—	—	—		—	5	3		—	—	—	
Carex aquatilis	50	60	75		—	50	10		95	10	70	
Carex rostrata	20	5	2		2	10	2		—	65	15	
Carex sp.	—	—	—		—	3	—		—	—	—	
Deschampsia caespitosa	15	5	10		15	15	10		5	15	15	
Epilobium sp.	—	—	—		—	—	5		2	1	1	
Hordeum brachyantherum	—	2	4		3	—	—		—	—	—	
Juncus arcticus	—	5	8		35	—	—		—	—	—	
Pentaphylloides floribunda	1	30	15		—	15	5		—	—	—	
Potentilla sp.	—	—	2		2	—	—		—	—	—	
Salix planifolia	5	3	—		3	5	30		—	—	45	
Unidentified forbs and grasses	1	3	3	Avg.	—	—		Avg.	—	—	2	Avg.
Total Cover	92	113	119	= 108	62	103	75	= 80	102	91	146	= 113
Open Water	5	—	—		—	20	—		—	20	25	

[a]These wetlands are formed over peat saturated during the entire year.
[b]Quadrat size was 1 × 2 m.

Source: Von Loh, 1991.

TABLE 8-4 Percent Cover of Reclaimed Area on Mineral Soils, South of Reclaimed Channel

	Transect 1			Transect 2			Transect 3		
Species	Quadrat: a	b		Quadrat: a	b		Quadrat: a	b	
Achillea millefolium	—	—		—	1		—	—	
Agropyron sp.	10	2		1	12		20	10	
Antennaria sp.	—	3		—	—		—	—	
Artemisia frigida	4	5		—	—		—	—	
Chaenactis douglasi	—	—		—	—		5	—	
Cirsium sp.	2	—		1	—		—	—	
Epilobium sp.	—	—		2	2		—	2	
Festuca arizonica	5	5		2	5		1	—	
Hordeum brachyantherum	5	—		—	—		—	—	
Juncus arcticus	2	2		1	1		—	—	
Phleum alpinum	—	—		—	2				
Phacelia heterophylla	—	2		—	—		—	—	
Poa sp.	—	2		—	—		2	3	
Polygonum aviculare	2	1		—	—		—	—	
Potentilla sp.	—	4		—	2		—	—	
Unidentified forbs and grasses	2	—	Avg.	7	5	Avg.	2	3	Avg.
Total Cover	32	26	29	14	30	22	30	8	24
Litter	10	30		5	20		2	2	
Bare ground	58	44		81	50		68	80	

Source: Von Loh, 1991.

spreading rapidly across this area. Existing portions of the transect were strongly influenced by the taller, more mature willow and bog birch shrubs, which shade out much of the herbaceous cover that otherwise might have developed. The species composition was similar on both sides of Hoosier Creek at this location.

A portion of the restored charcoal dump lies upslope of the beaver pond, which still influences the groundwater table, but the area is also fed by springs flowing under the road on the northern edge. Subsequently, the entire area was saturated year round. Growth from willow brush layer cuttings is most pronounced at this site, averaging approximately 40% aerial cover in the reclaimed portion. Herbaceous species, particularly sedges, did little colonizing in the reclaimed area, perhaps the result of fairly severe frost heave, noted in the early part of the growing season.

Wildlife use through the final year of monitoring continued to increase sitewide, particularly waterfowl breeding and brood rearing. Elk continued to graze and browse across the enclosure but no longer pulled willow cuttings from the peat because the root systems were well developed. It appeared that a good population of brook trout was present in Hoosier Creek and in the ponds behind the beaver dams. There was an increase in the numbers of macroinvertebrates that were observed, although no attempt was made to determine the species composition.

Despite the alteration to the site brought about by beaver and elk, these animals were not controlled. Beaver were allowed to develop dams and to forage around the resulting ponds. The relatively low fence surrounding the mitigation site was easily crossed by elk, and they continued to browse freely on the site. No non-native, invasive plant species appeared during the 3-year monitoring program. Moreover, because the hydraulic control structure on Hoosier Creek was fixed and could not be easily adjusted, water levels were not manipulated except by beaver. Consequently no management programs were employed.

The on-site mitigation was not as carefully monitored. The plantings suffered the same predation, but the results appeared, from casual inspection, to be satisfactory. This observation undoubtedly was shared by the project manager for the Corps, for he ruled that the permit conditions had been met.

CONCLUSIONS

The formulation, execution, and results of this mitigation effort were characterized by a good understanding of purpose and intent by the various participants, confined mainly to the developing agency, the Colorado Department of Transportation, and the regulators, the FWS and the Corps. Although there was written documentation, the work proceeded, in large, on a "handshake" agreement. Many standard elements of a mitigation plan, such as the development of clear goals and the establishment of success criteria, were not articulated nor were construction drawings and specifications formulated for the off-site mitigation. In the end, representatives of the agencies involved in reviewing and approving the 404 permit all agreed that both the on-site and off-site mitigation efforts were successful (Fig. 8-8).

Many of the individuals representing those agencies agreed that a large part of the success was attributable to the stewardship of Ed Demming with the Colorado Department of Transportation. He not only took a great interest in the restoration, but also gave considerable time and thought to the process. And he was not alone in his efforts. Many others at the Department of Transportation as well as with the regulatory agencies took a personal interest. The project manager for the Corps visited the site on numerous occasions, closely following the progress of restoration.

Cooper's restoration plan, as simple as it was, established the course of restoration. His reliance on nature's ability to self-design was one of the keys to success. With a minimum of plant introduction—save for the willow sprigs—the propagation of the site by a diverse, robust plant community was accomplished by natural forces: streamflow, wind, beaver, and other wildlife. First, however, as recommended by Cooper, the hydrologic system was put in place. Having achieved suitable subsurface and surface water regimes, the plants and animals worked together to create a successful restoration project.

Figure 8-8 Restored sedge meadow with shrub carr fringe. Photograph by Kristine Meiring. Photo also appears in color insert.

Plants from the southern, extant wetland began to invade the newly created land form on the north side of the creek. Upstream of the constructed hydraulic control, beaver found suitable habitat to build dams, which further modulated streamflow through the reach. The sustained flows and higher groundwater levels provided by the beaver dams undoubtedly favored the development of the sedge community. The beaver dams reduced the effects of high velocities and provided water during the drier periods of the year, which would not have been available in the absence of their dams and in the presence of the formerly eroded channel.

Although a reasonably extensive monitoring program was designed and executed, the results did not lead to, and perhaps did not require, any intervention in the development of the new habitat. Representatives of the regulatory agencies and the implementing scientists seemed satisfied with the emerging results, including the sedge meadow and the diminished willow carr. Consequently, other than minor repairs to the controlling rock weir, there was no adaptive management.

ACKNOWLEDGMENTS

The success of this project was directly related to the interest and commitment of the various participants. These participants ranged from the primary regulator, the Army Corps of Engineers, through the construction agency, the Colorado Department of Transportation, to the scientific community, as represented by Dr. David Cooper, Research Scientist at Colorado State University, Fort Collins. Much of the material and insights to the project were provided to the authors by Kristine Meiring, with the Colorado Department of Transportation. She prepared the "Wetland Opinions," helped monitor the site, and shepherded the project along the trajectory to success. She documented the project's progress with photographs and videos and made public presentations. Ed Demming, the project engineer, was most responsible for the successful restoration of the hydrology at the Hoosier Creek site. He designed the control structure at the railroad embankment and the backfilling of the eroded channel. On a daily basis, he oversaw the construction of these restoration elements. The concept of the project was the work of Cooper. He evaluated the site and recommended that it be used for mitigation, he installed the first groundwater wells, and he highlighted the importance of hydrology in the restoration effort. He also foresaw the role that beaver might play in the restoration. Despite his intention that the site be more heavily vegetated by willow, in the end, he was determined to promote the wetland features of the site, regardless of the species composition. Terry McKee, project engineer with the Corps, took an extraordinary interest in the work, carefully monitoring its progress. He evaluated the progress and ultimately approved the end results. Bill Noonan, a biologist with the U.S. FWS, and Jim Von Loh, a botanist with the Colorado Department of

Transportation, both participated in field studies and in project review. Harold Warren, a resident of Grant, Colorado, provided an historical perspective.

REFERENCES

Bowman, F., *Letter to District Engineer, Omaha District, U.S. Army Corps of Engineers*, Park County Planning Commission, Fairplay, CO, 1988.

Butler, V. W., *Letter to District Engineer, Omaha District, U.S. Army Corps of Engineers*, Upper South Platte Water Conservancy District, Fairplay, CO, 1988.

Cooper, D. J., *A Handbook of Wetland Plants of the Rocky Mountain Region*, U.S. Environmental Protection Agency, Region 8, Denver, CO, 1989.

Cooper, D. J., Personal Correspondence, 1998.

Cooper, D. J., *Wetland Restoration in the Hoosier Creek Valley: Colorado Front Range*, Colorado Department of Highways, District 1, Dumont, CO, 1988.

Cooper, D. J., *An Evaluation of the Effects of Peat Mining on Wetlands in Park County, Colorado*, Park County, Colorado, Fairplay, CO, 1990.

Dahl, T. E., Wetland Losses in the United States: 1780s to 1980s, U.S. Department of Interior, Fish and Wildlife Service, Washington, DC, 1990.

Hand, O. D. and D. Angulski, *Archaeological Survey and Test Excavations of the Webster-NE Wetland Mitigation Site, Park County, Colorado*, Colorado Department of Highways, Denver, CO, 1988.

McOllough, P. R., *Letter to Shawnee Water Consumer's Association*, Colorado Department of Highways, Aurora, CO, 1988.

Meiring, K. J., J. D. Von Loh, and D. J. Cooper, *Hoosier Creek Wetland Restoration*, Colorado Department of Transportation, Denver, CO, 1991.

Opdycke, J. D., *Letter to Chief, Regulatory Branch, Omaha District, U.S. Army Corps of Engineers*, U.S. Fish and Wildlife Service, Grand Junction, CO, 1988.

Painter, E. F., *Letter to District Engineer, Omaha District, U.S. Army Corps of Engineers*, Bailey Water and Sanitation District, Bailey, CO, 1988.

Price, A. B. and A. E. Amen, *Soil Survey of Golden Area, Colorado*, United States Department of Agriculture, Soil Conservation Service, Washington, DC, 1984.

Tolpo, V., Letter to District Engineer, Omaha District, U.S. Army Corps of Engineers, Shawnee Water Consumer's Association, Shawnee, CO, 1988.

Von Loh, J. D., *1991 Hoosier Creek Wetland Restoration Site Monitoring*, Colorado Department of Transportation, Denver, CO, 1991.

Von Loh, J. D., *1990 Hoosier Creek Wetland Restoration Site Monitoring*, Colorado Department of Transportation, Denver, CO, 1990.

U.S. Geological Survey, *Water Year, 1995*, Water Resources Data Report, Denver, CO, 1996.

Weber, D., *Letter to District Engineer, Omaha District, U.S. Army Corps of Engineers*, Colorado Department of Natural Resources, Denver, CO, 1988.

Weber, W. A., *Rocky Mountain Flora*, University Press of Colorado, Niwot, CO, 1976.

CHAPTER 9

KACHITULI OXBOW

The Sacramento River historically formed a wide, verdant valley in north-central California (Fig. 9-1). Tule (*Scirpus acutus*) marshes and riparian habitat dominated the flat floodplain lands stretching from San Francisco Bay to the foothills of the Sierras. The tule marshes stored the large volumes of water that were discharged from the snowpack high in the Sierra Mountains. They also provided food and fiber for the sparse populations of native Americans prior to European settlement. The long, narrow bands of riparian habitat, established on natural levees bordering the Sacramento River channels, provided homes for diverse, productive wildlife populations. From these landscapes, settlers created a highly productive agricultural economy. As this economy grew, so did the towns and cities along the course of the river, and this growth continues today. Many of the agricultural communities and landscapes are giving way to urbanization. A case in point can be found in southeast Yolo County.

Yolo County is largely agricultural. However, its southeastern corner, across the river from and abutting the city of Sacramento, is urbanized and growing (Fig. 9-2). As a result, Yolo County and the communities surrounding Sacramento are seeking development to pay for infrastructure and to secure jobs for their residents. With these interests in mind, in 1985, the county and a newly forming municipality, the city of West Sacramento, along with a developer, pursued the acquisition of surplus federal lands and the redevelopment of these and adjacent properties. These land holdings are at the confluence of the Sacramento and American Rivers (Fig. 9-2) and are partially protected by a right-bank levee.

The combined properties total approximately 280 acres, falling both on the river side and the protected side of the Sacramento River levee. In the lee of

Figure 9-1 Project location.

the levee, the developer proposed to construct 1,881 dwelling units, a 500-room hotel, a large convention center, business and professional offices, a retail commercial center, a 700-person arena, and a yacht club. The developer also proposed to redesign an existing golf course and to provide open space. The estimated cost of development was $300 million (Levin, 1985)—Lighthouse Marina was the project's given name.

The project had its origin with the acquisition, by Yolo County, of approximately 18 acres of surplus state and federal land. This land originally had been acquired by and for the U.S. Army Corps of Engineers (the Corps) to support levee construction along the Sacramento River. Having completed the levees, the Corps no longer needed the land. Because no federal or state agency wanted the land [the U.S. Fish and Wildlife Service (FWS) and the California Department of Fish and Game (DFG) declined to take it (Green, 1985)], the county

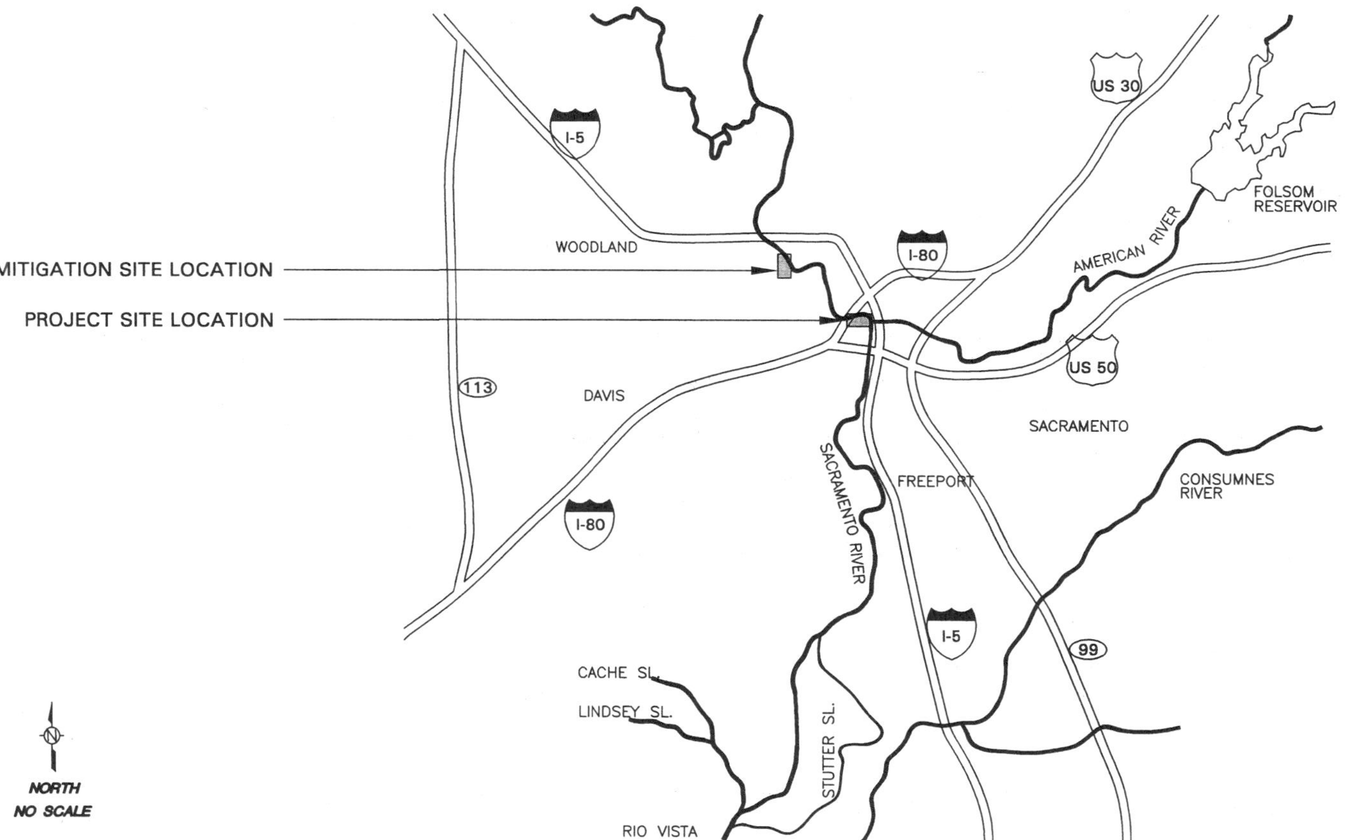

Figure 9-2 Sacramento metropolitan area showing project and mitigation sites.

offered to buy it. This land was then combined with several other developed parcels, including three existing marinas and a golf course. Approximately 70% of the area, however, was undeveloped, formerly agricultural fields. Weedy vegetation and bare soil dominated these fields. The remnants of a gravel pit remained as a scar on the former floodplain of the Sacramento River.

The development plan for the assembled parcels of land was quite ambitious. The residential portion of the development was intended for high and moderate income housing. The golf course was to be improved. The three small marinas were to be consolidated into one much larger facility including a yacht club. In the end, only 10% of the site was to remain as open space.

In the fall of 1985, the project concept was presented at a public hearing held by the Community Development Agency of Yolo County. From the very beginning, and throughout its development and implementation stages, controversy plagued the project. On the one hand, officials of Yolo County were very supportive of the project in the belief that it would help revitalize a blighted area of unincorporated Yolo County, soon to be incorporated as the city of West Sacramento. On the other hand, a number of environmental organizations and state and federal resource agencies opposed the project. In support of the project, the county's development team presented a number of arguments in the Draft Environmental Impact Report/Environment Impact Statement (EIR/EIS) (EDAW Inc., 1986):

- Community Enhancement—The fundamental benefit of the project is its capacity to begin reversing the deterioration and blight in East Yolo. The project will reintroduce a very high-quality development market into the East Yolo community, enhancing the riverfront itself as well as anchoring revitalization of the rest of East Yolo, commencing with the central business district . . . new business, jobs, infrastructure, taxes, and public facilities within the project area—and indirect benefits—new businesses elsewhere in East Yolo, new jobs, taxes, etc., produced by a "ripple effect" spreading throughout the community.
- Creation of Jobs—A principal benefit of the Lighthouse Marina Project is the creation of new employment opportunities in East Yolo . . . as of the 1980 census, there were 1,341 unemployed residents in East Yolo, which constitutes 11.5 percent of the labor force, as against a 7.9 percent unemployment rate in the greater Sacramento metropolitan statistical area.
- Low-Income Housing—The project provides a major opportunity to upgrade low-income housing through absorption of low-income tenants into the existing and new housing supply.
- Parks—The project provides significant public benefit with regard to enhanced open space and park use. The county development team and the project developer have agreed in concept to a design program for a linear riverfront park to run the full length of the project. This will provide public access to the entire riverfront area. At present, the entire riverfront is private property except the county boat ramp and Corps property.

- Security and Fire Protection—The project will provide a major benefit with regard to the criminal activities currently occurring on the site, and the resulting problems of the police and fire protection services. These problems are caused by a dense underbrush. Recent criminal activity within this area has included murder, arson, rape, assault without deadly weapons, robbery, cultivation of marijuana, dope trafficking, disorderly conduct, public drunkenness, and many other types of crimes.
- Infrastructure—The project will provide major public benefits with regard to the infrastructure required to service it. . . . The developers must provide for sewer, water, and storm drainage services, which will involve major upgrades of those present systems. Even though the need for the upgrades is produced by the project, the new capacities in the upgraded system will generally benefit the East Yolo area.

To a lesser extent, the economic interest of the General Services Administration (acting on behalf of the Corps) and the California Department of Transportation was being served by the project and its developers. Both agencies owned surplus land that would be purchased by Yolo County and, ultimately, by the developer. Although the sale of their properties did not represent a significant financial stake, both agencies were clearly interested in pursuing the sale.

The opposition to the project was formidable. The FWS was adamantly opposed to it because of the loss of riparian habitat and, in particular, the loss of habitat for the Valley elderberry longhorn beetle (VELB) (*Desmocrous californicus dimorphus*), a federally listed, threatened species. The FWS and the DFG also were concerned about state-listed species: Swainson's hawk (*Buteo swainsoni*), Western yellow-billed cuckoo (*Coccyzus americanus occidentalis*), and the Sacramento River winter-run chinook salmon (*Oncorhynchus tshawytscha*). The EPA raised its voice over the loss of wetlands and shaded riverine habitat and the effects of the project on the water quality of the Sacramento River. The Corps had regulatory authority over the project under two separate legislative mandates (Section 10 of the Rivers and Harbors Act of 1899 and Section 404 of the Federal Water Pollution Control Act of 1972) and was responsible for issuing the appropriate permits. Under the first act, the Corps regulates development on the river side of the levee. The second act gives the Corps authority over wetlands on both sides of the levee. In the end, the Corps had the responsibility to decide whether or not the public interest was served by the project.

The FWS was joined in its opposition by the DFG. The DFG looked unfavorably upon any development on the river side of the levee and expressed concern about two active Swainson's hawk nests near the project, which had been used in previous years. The DFG also decried the potential loss of fish and wildlife habitat along the riparian corridor, including the elderberries. Elderberry shrubs serve as the host plant for the federally listed, threatened species VELB. These agencies were joined by numerous environmental groups including, but not limited to, Defenders of Wildlife, Davis Audubon Society, American Fisheries Society, and The Sacramento River Preservation Trust.

In April of 1985, the developer and Yolo County jointly submitted an application to the Corps to fill wetlands and impact riparian habitat on the property. Following an extensive review of the application, on August 16, 1985, the Commander of the Sacramento District of the Corps concluded that: ". . . the proposed project may have a significant effect on the environment, and an Environmental Impact Report is required" (U.S. Army Corps of Engineers, 1985). This decision was followed by a public notice, on August 29, 1985, requesting comments on the scope of the joint EIR/EIS. This started, in earnest, the public debate.

The conservation groups, individually and in concert, urged the Corps to deny the Section 10 and Section 404 permits. They argued that precious little shaded riverine aquatic habitat remained along the Sacramento River. The Defenders of Wildlife (Spotts, 1986) pointed out that "The Sacramento River provides some of the best remaining riparian habitat in northern California, despite a reduction in habitat from an estimated 800,000 acres in 1848 to approximately 12,000 acres today." The conservation groups wrote numerous letters to each of the cognizant federal and state agencies arguing for not selling the surplus lands to the county, denying the permits, or requiring that the developer set aside the riparian and elderberry savanna habitats. Each agency was reminded of its regulatory mandates starting with the General Services Administration, which was responsible for the property and its disposal. This agency was repeatedly reminded that it had responsibility under Executive Order No. 11988 for floodplain management and Executive Order No. 11990 for wetlands protection. The opponents urged the Director of Real Estate Sales, Region 9, of the General Services Administration, not to transfer the property to Yolo County unless there was a legal finding that the transfer was fully in compliance with the executive orders.

The Draft EIR/EIS was submitted in early September, 1986 (EDAW, Inc., 1986), and the public notice of the draft was issued later in the month. In this document, the proponents of the development project argued that Executive Order No. 11988 governing floodplains had been met and that, equally so, Executive Order No. 11990 governing minimization of wetland losses had been satisfied. Responses to the opponents of the project were addressed in the Final EIR/EIS (EDAW, Inc., 1986). Also, a conceptual mitigation plan was prepared as part of the process.

Anticipating that the Corps would issue the permits, in November of 1986, nine environmental groups joined together to express their opinions about the Final EIR/EIS and the pending action of the Corps. They stated that "None of the undersigned groups are fundamentally opposed to the project. Rather it is in our interest to see that the impacts to riparian habitat, fisheries, and endangered species are adequately mitigated. To demonstrate our desire to maintain a positive approach to this project [Lighthouse Marina] we have prepared the attached guidelines for riparian habitat mitigation. These guidelines may be considered as minimum specifications for the detailed mitigation plan'' (Sacramento, 1986). These guidelines were ultimately used in forming the basis of the mitigation program.

On February 17, 1987, the Corps' District Engineer issued the combined permit, number 9051, allowing the project to proceed.

The next several years involved designing the development project and the mitigation program and securing the necessary financial backing. On December 18, 1990, the developer submitted to the Corps a request to revise the mitigation requirements, the reason being that the amount of riparian habitat and elderberry savanna to be destroyed by the project was reduced. In Permit No. 9051, 145 acres of mitigation was required for the loss of riparian habitat and 48 acres for the loss of elderberry savanna. If revised, the new area for riparian habitat would have been approximately 110 acres and that for elderberry savanna habitat, 45 acres. The proposed change was denied by the Corps. Consultants for the developer proceeded to identify suitable properties for the original mitigation conditions. A site for the riparian habitat was selected in 1990 and a plan submitted to the Corps for approval. The wetland portion of the mitigation plan was contained within the plan for riparian habitat. The selected mitigation site was named Kachituli Oxbow, after an Indian village that once existed near the property (Kroeber, 1925).

ENVIRONMENTAL AND SOCIAL SETTINGS

The proposed project, Lighthouse Marina, and the mitigation site, the Kachituli Oxbow, are located near the City of Sacramento in Yolo County. The Sacramento River forms the northern boundary of the Lighthouse Marina project, and the project's northeast corner lies at the confluence of the Sacramento and American Rivers. The mitigation site is 7 miles upstream from West Sacramento on the west bank of the Sacramento River, in Yolo County.

Yolo County has a broad range of landscapes, varying from hilly to steep mountainous land in the California Coastal Ranges, to the broad, flat valley of the Sacramento River. The Soil Conservation Service (SCS, which is now known as the National Resources Conservation Service) estimates that two-thirds of the county's 1,034 square mile area is in the Sacramento Valley between the coastal ranges and the Sacramento River. The elevation difference between the valley and the coastal mountains is on the order of 2,000 feet. The valley floor, or the central part of Yolo County, starts at an elevation of 16 feet above mean sea level (fmsl) at the southeastern edge and rises to 32 feet at the northern boundary.

Yolo County lies within the great valley province (Andrews, 1972). This province includes the Sacramento and San Joaquin valleys and a small part of the coastal ranges, which are a series of mountains running parallel to the California coast. Yolo County consists of five separate geomorphic units. These are (1) floodplains and natural levees; (2) flood basins; (3) low alluvial plains, fans, and terraces; (4) low hills and dissected terraces; and (5) uplands of the coastal ranges. The project and mitigation site are located in floodplains and natural levees.

The climate is characterized as Mediterranean, with warm, dry summers and cool, moist winters. The average daily temperature is approximately 62°F, with the average daily maximum being 76°F and the average daily minimum 49°F. There is a temperature gradient across the county from southwest to northeast that is influenced by air currents flowing eastward from the Pacific Ocean, the cooler temperatures being near the ocean. Owing to the moderate climate of the county, the growing season is quite long. With a 50% certainty, the growing season will last from April 10 through November 12, over 216 days.

Approximately 22 inches of precipitation falls on the county. The vast majority of this moisture occurs as rainfall, less than 1% as snowfall. Approximately 90% of precipitation occurs during the months of November through April.

The water resources of the county depend on Cache Creek, Putah Creek, the Sacramento River, and the Glenn–Colusa Canal, which imports water from the Sierras and the southern Cascades. Based on the U.S. Geological Survey's Stream Gauging Station at Sacramento, the mean yield of the Sacramento River is 15 inches. The yield, of course, is little influenced by the climatic conditions of the county in that the Sacramento River and its tributaries rise in the Sierras and a substantial portion of the runoff water is snowmelt. On the other hand, Cache Creek reflects Yolo County's hydrologic conditions. The yield of this stream is less than 3 inches. Given the 21 inches of precipitation, runoff is less than 14% of precipitation. Evaporation rates are quite high owing to the long, warm, dry periods of the year. As a result, most crops are produced with the aid of irrigation.

The soils of Yolo County are composed of twelve associations. These associations are aggregated into two groups: alluvial fan or flood basin soils and upland and terrace soils. Seven of the associations belong to the first grouping and cover 63% of the county. The remaining five associations account for 37% of the county. The first category contains the following associations: (1) Yolo–Brentwood, (2) Rincon–Marvin–Tehama, (3) Capay–Clear Lake, (4) Sycamore–Tyndall, (5) Sacramento, (6) Willows–Pescadero, and (7) Capay–Sacramento. These soils are considered to range from well-drained to poorly drained and are positioned on alluvial fans, basin rims and terraces, and in basins. This grouping, and in particular the Sycamore–Tyndall association, dominates both the project and mitigation sites. The other categories of associations are made up of soils that are excessively drained to well-drained. These associations include (1) Corning–Hillgate; (2) Sher–Horn–Balcom; (3) Dibble–Millsholm; (4) Positas; and (5) Glockwin.

Most of these soils are better drained today than in the past, owing to subsurface drainage and the control of flooding brought about by the extensive levee systems on the main stem of the Sacramento River and its tributaries.

Also, large areas of native soils have been buried by the debris produced by hydraulic mining in the Sierra Nevada foothills (Andrews, 1972).

In particular, the Silty Sycamore series of the Sycamore–Tyndall association is found on floodplains and natural levees. These soils have existed long enough to accumulate some organic matter, giving them a darkened surface horizon (Andrews, 1972).

Four soils exist on the project site: Lang sandy loam; Lang sandy loam, deep; Sycamore silt loam; and Tyndall very fine sandy loam. With the exception of Lang sandy loam, these soils are nonhydric. In the case of Lang sandy loam, some soil mapping unit components can be hydric; however, test pits excavated during the wetlands surveys did not find these hydric components present. Eighty percent of the Lang sandy loam lies in the area to the riverside of the levee. Only a small portion of the site in the southeast corner is underlain by Lang sandy loam.

The seasonal high groundwater for the area landward of the levee, based on an examination of soil character and standing groundwater levels, is 3 to 5 feet below the land's surface. The soils, having been deposited by fluvial processes, are heterogeneous with strata of silt and clay. The primary sources of groundwater are the Sacramento River, which recharges the more permeable fine sand and silty clay lenses near the river, and upslope charging of acquifers from the western drainages. A primary source of water on the Kachituli site is infiltration and percolation of precipitation. Groundwater levels fluctuate with river levels.

The plant communities found on the project site conformed to the various man-made and natural landscapes. These landscapes include the levee, golf course, small residential developments, farmed fields, and remnant floodplain. Accordingly, the plant communities were identified and mapped (Fig. 9-3) as follows:

- Annual grassland/ruderal forb.
- Elderberry savanna.
- Agricultural.
- Riparian woodland containing the following subcategories:
    - Upstream section
    - Middle section
    - Downstream section
    - Young woodland and ditch woodland
    - Oak woodland
    - Ornamental.

The largest plant community was making up 32% of the site. The next largest was annual grassland/ruderal forb, accounting for 31% of the site. Riparian woodland, covering all five subcategories, made up approximately 19% of this site. The smallest communities were elderberry savanna (1%) and oak woodland (2%).

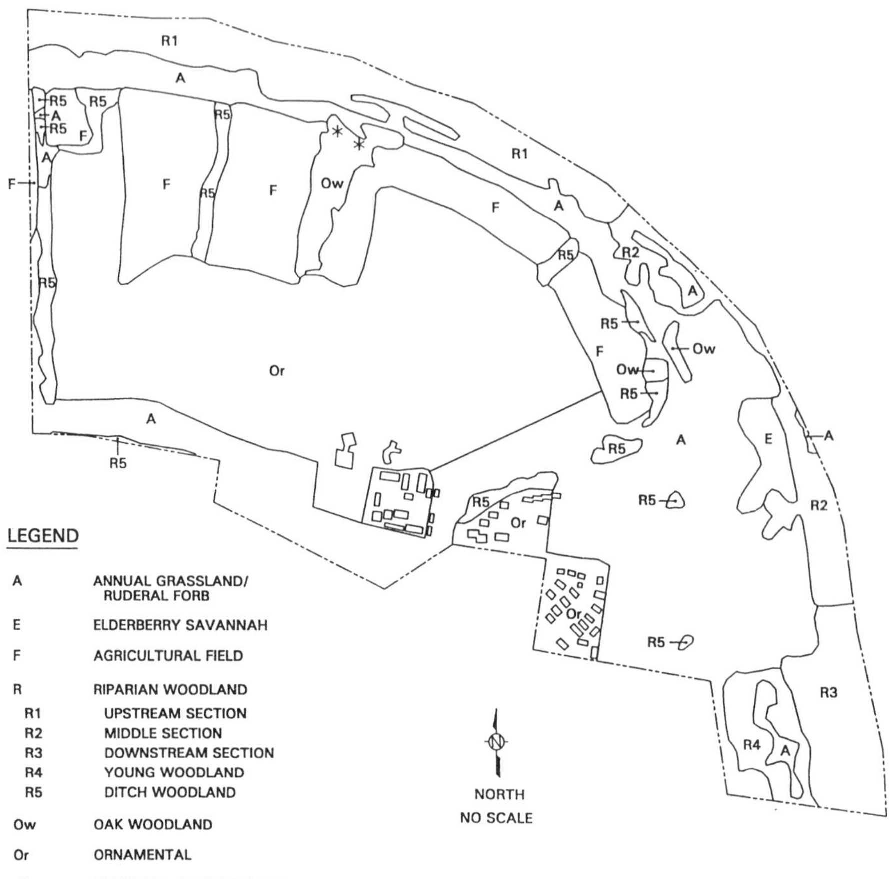

Figure 9-3 Land cover and plant communities.

Prior to European settlement, the Sacramento River flowed between low, natural levees bordered by vast expanses of tule marshes. These marshes extended from the Sacramento–San Joaquin Islands to Colusa, California, a distance of 100 miles. Laterally, the marshes extended 2 to 3 miles. Tule lowlands and marshes blocked cross-valley travel. During the spring, the season of high water, oaks and cottonwoods standing on the natural levees were visible above the water level. The water spread out across the marshes, extending for miles on either side of the Sacramento River (Kelley, 1989). The tule marshes and riparian habitat were the dominant landscape along the Sacramento River. The natural levees bordering the river supported woody vegetation whereas the adjoining flood basins supported a variety of emergent vegetation. These wetland vegetation forms were supported out to the 100-year flood limit (Katibah, 1984).

The valley was not explored by Europeans until 1821, when Luis Argüello arrived in search of a mission site (Hansen, 1967). Fur trappers arrived a little later the same year on a mission for the Hudson Bay Company. As European settlement occurred in the 1840s, the tules were cut and burned to allow passage across the Sacramento Valley. Because of the desire to develop the floodplains of the Sacramento River and the tremendous floods that inundated these lands during the winter rainy period, the settlers began to construct levees along the river and its tributaries in 1867. These man-made structures replaced or built upon the natural levees and reduced the width of the river from miles to feet from the San Joaquin Delta to Colusa. The loss of the tule marshes, along with wetland losses along other major rivers in California, has earned California top ranking as the state with the greatest percentage of wetland losses (Dahl, 1990). Today, in the vicinity of the project and including the project itself, levees enclose the Sacramento, Feather, and American Rivers. The tule marshes are almost gone.

Although early settlement and the development of agriculture in the Sacramento River Valley rapidly expanded during the late 1800s, Yolo County remained agricultural throughout this century. The urbanization that has occurred has come mostly from the cities of Davis and Woodland. In 1900, the population of Yolo County was approximately 14,000 or, expressed a little differently, the population density was 13 persons per square mile (Fig. 9-4). By 1940, the population density had doubled to approximately 26 persons per square mile. It wasn't until the late 1950s that the population had again doubled and then, in the 1980s, it began to increase more rapidly. By 1990, the population had reached 141,000 or 136 persons per square mile, six times what it was in 1940. On the other hand, the more urban county to the east, Sacramento County, saw steady population increases in the 1900s. In 1900 this county had a resident population of approximately 46,000. By 1990 the population stood at over 1 million. The population density in this county is more than 1,000 persons per square mile, about 10 times that of Yolo County.

Today, the major industry of Yolo County is still agriculture, with related packing and canning industries. The principal field crops are sugar beets, alfalfa, rice, sorghum grain, barley, corn, wheat, and safflower. Tomatoes are an important export crop, as are almonds and fruits. More than 77% of the land area in Yolo County is farmed.

THE PROJECT AND ITS WETLAND IMPACTS

The proposed project was to be constructed on the west-bank low-stream terrace and former floodplain of the Sacramento River. The topography is relatively flat, ranging in elevation between 15 and 20 feet above mean sea level (fmsl). The highest point on the property is the top of the levee, at 40 fmsl. Several borrow sites on the property extended below 15 fmsl.

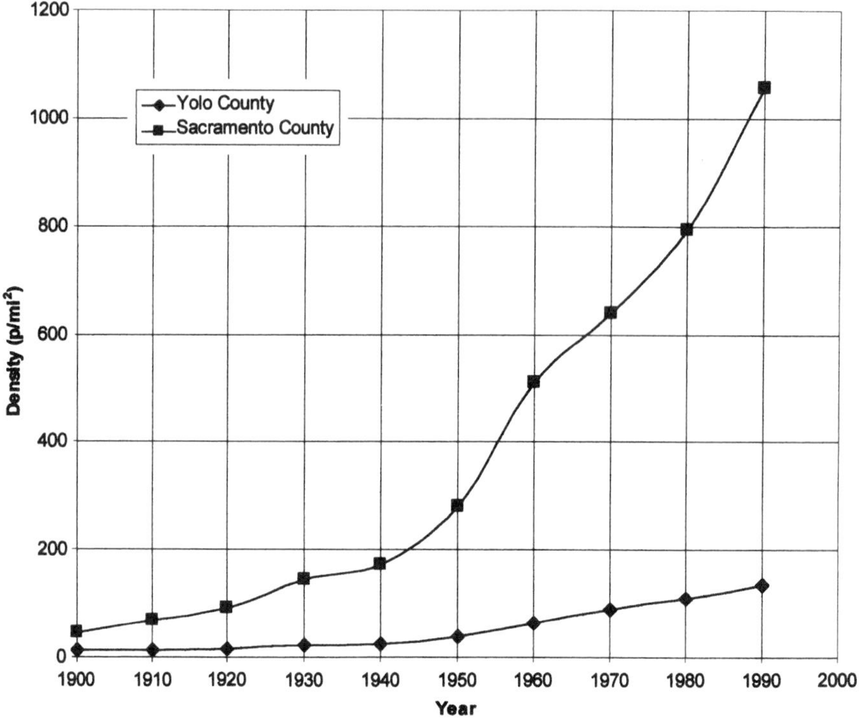

Figure 9-4 Population densities of Yolo and Sacramento Counties, California.

Approximately 75% of the property (210 acres) is enclosed by a levee, which runs from the northwest corner in an arcing fashion to the southeast corner. The setback of the levee from the Sacramento River is approximately 200 feet. The levee protects the enclosed property from the 100-year flood (Fig. 9-5). Because the levee is a part of the Sacramento River Flood Control Project, undertaken by the federal government and the State of California, the Corps and three state agencies (State Reclamation Board, Department of Water Resources, and State Lands Commission) have vested interests. Each agency acted on its authority in reviewing the project.

Throughout development of the project and the public debate, little attention was given to the wetlands on the property. The Corps invoked Section 404 authority, but the wetland characteristics of the site were all but ignored. The main focus of the inter-agency debates was the riparian and upland habitat associated with threatened and endangered federal- and state-listed species. A wetland survey, however, was conducted and the results became a part of the draft EIR/EIS (EDAW, Inc., 1986). Curiously, survey data were collected only on the river side of the levee in the Lang sandy loam soils (Fig. 9-6). The survey resulted in a mixed conclusion. Employing the criteria set out by the FWS (Cowardin et al., 1979), the plant communities at the sample points in-

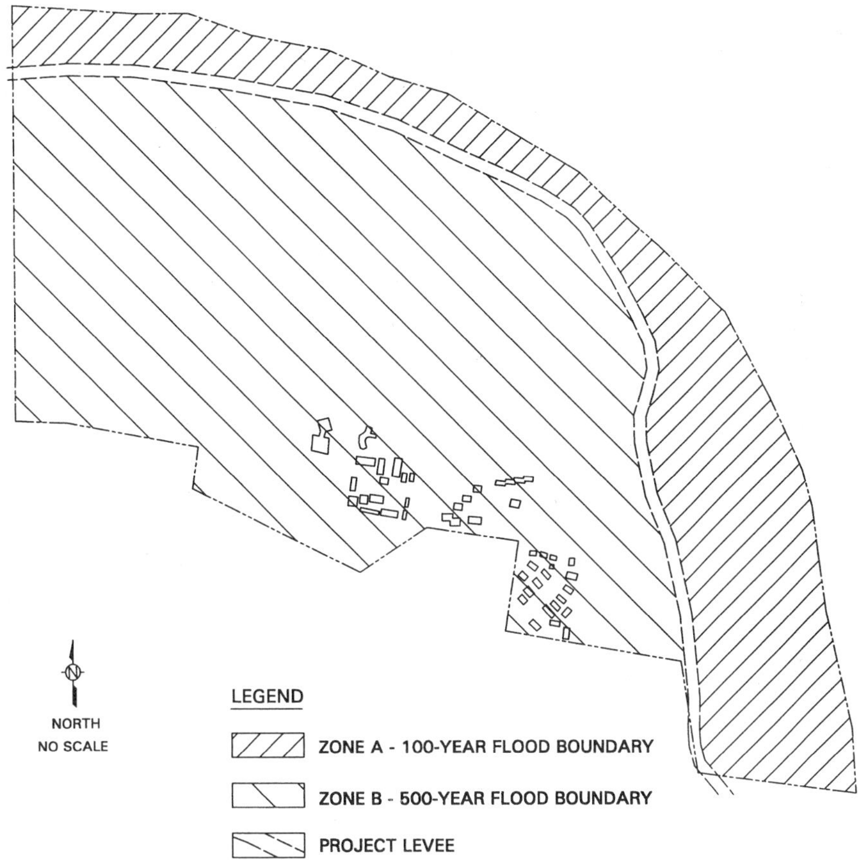

Figure 9-5 Floodplain boundaries.

dicated wetlands. Using the Corps criteria set out in the 1985 manual (U.S. Army Corps of Engineers, 1982), none of the sample points was classed as wetland. None of the swales or the borrow pits on the landward side of the levee was sampled.

In the end, the wetlands were delineated by fiat rather than fact. The "Record of Decision" regarding the 404 permit defined the wetlands on the project site as all those areas "lying below the 18-foot elevation contour." This wetland definition was agreed to by all of the parties concerned, and there appeared to be little or no debate. The wetland comprises 45 acres. Again, the wetlands were not the focus of the controversy or the mitigation plan. Although they were fully mitigated, the riparian habitat and the VELB habitat drove the mitigation plan.

The permit contained a number of conditions, including the following:

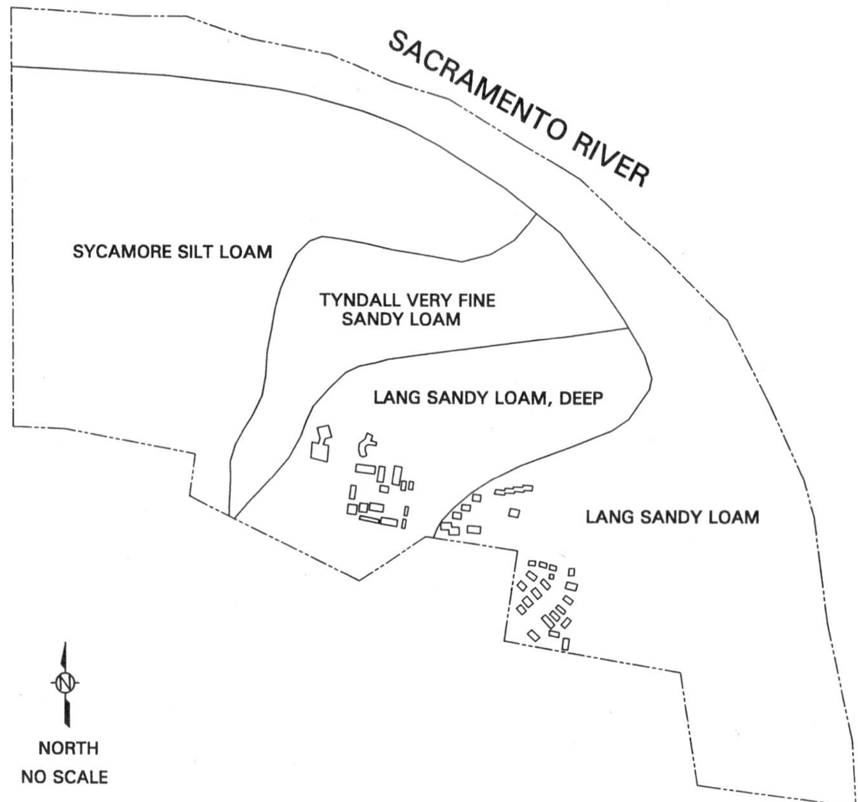

Figure 9-6 Soil mapping units of the project site. Source: Andrews (1972).

- The permittee shall survey and clearly mark the 18-foot elevation contour within the project boundaries. Project contractors will be notified that no fill material will be placed below the 18-foot contour, other than that allowed within this permit for backfill and bank protection, without obtaining an additional Corps permit. Upon completion of the above survey the permittee will notify the Sacramento District and allow the opportunity for inspection prior to construction. The permittee shall submit final project plans, which show the survey 18-foot contour in relation to the project layout, to the District Engineer 30 days prior to construction (the area below the 18-foot contour was considered to be wetland).
- The District Engineer intends that the impacts of the Lighthouse Marina project to the valuable riparian habitat on the water side of the federal project levees shall be compensated. Thus the District Engineer requires the permittee to acquire 145 acres of open space or agricultural land suitable for development and/or restoration of riparian habitat. These 145 acres are subject to the following conditions:

- The parcel shall be approved by the District Engineer and acquired 30 days prior to construction, or no more than 1 year after the issuance of this permit, whichever is first.
- If the land is acquired in more than one parcel, the parcels must be contiguous or neighboring.
- All mitigation lands must be located along the Sacramento River, and unless shown to be infeasible, in Yolo County.
- The total mitigation land(s) must be dedicated in perpetuity for wildlife habitat through conditions, covenants, and restrictions (CC&Rs) recorded with the title to these lands. No improvements may be made nor operations conducted in the mitigation lands whose purpose is not primarily to further the status of the land as wildlife habitat.
- The permittee shall monitor and maintain the restoration and/or developed habitat on the mitigation lands for five years following completion of the habitat development and/or restoration work. . . . If at the end of five (5) years the habitat restoration and/or development is not fully successful, the permittee shall provide, as prescribed by the District Engineer, such additional reasonable and practical mitigation that is necessary to achieve the District Engineer's above-stated intent.
- The 145 acres of mitigation land may be reduced to a lesser amount if further analyses show such reduction to be appropriate following final design of the project. Any such analysis will be done by the Sacramento District Corps of Engineers.

Attached to the Corps' permit was the report from the FWS on the endangered species consultation. This report was appended and made a part of the permit. It contained additional mitigation requirements, specifically for the threatened VELB.

Approximately a year after the permit was issued, proponents of the Lighthouse Marina project submitted a more detailed plan for meeting the special conditions of the permit. This plan consisted of two parts: endangered species compensation and off-site riparian mitigation. The loss of VELB habitat was estimated at 16 acres. Given the required 3 to 1 mitigation ratio, mitigation was set at 48 acres, as specified in the special conditions of the 404 permit. On the other hand the riparian habitat losses were estimated to be 45.4 acres and the permit required the restoration of 145 acres. The total mitigation was tentatively set at 193 acres. The Corps recognized, however, that these acreages were not the final numbers. They allowed the proponents of the project to resurvey the site and make necessary adjustments to the development plan and then return to the Corps with the final numbers. This was done in a letter on December 18, 1990 (Corollo, 1990). In the letter, the 145 acres of riparian habitat mitigation was reduced to 108 acres and the VELB habitat was reduced from 48 to 40.5 acres. Despite the proponents' analysis and arguments, the permit conditions remained the same: a total of 193 acres suitable for off-site mitigation split between riparian and VELB habitats.

Within the geographic limits set out by the Corps, a number of potential off-site mitigation sites were identified. These sites were surveyed for soils, plants, and the needed modifications to meet the mitigation conditions. Although considerable effort was made to locate a single parcel that would accommodate the entire 193 acres of off-site mitigation, no such parcel could be found. Project proponents petitioned the Corps to divide the mitigation among an additional one or two sites. This was approved.

The main component of the mitigation plan involved the restoration, or creation, of riparian habitat. A suitable site was located 7 miles upstream on the west bank of the Sacramento River (Fig. 9-2). The 254-acre property was known as the Amen Ranch, in recent years used in the production of tomatoes (ECOS, 1989). The Corps approved the property for off-site mitigation, although a considerable amount of excavation would be required. The portion of the Amen Ranch allowed as mitigation for the Lighthouse Marina project included 110 acres suitable for mitigation of riparian habitat, of which 100 acres were on the land side of the levee and 10 acres on the river side. This left 83 acres of mitigation to be found elsewhere.

Further upstream, a suitable location was found for the restoration of the VELB habitat. Still, additional off-site acreage was required and this was gained through acquisition and restoration of riparian habitat in one of the sites used to develop the design criteria for the off-site mitigation work at the Amen Ranch property. This site was known as Mary Lake. Although creation and restoration work were required for the latter two mitigation parcels, and these efforts were monitored, only the Amen property and the associated mitigation are discussed here, because this property was used for wetland and riparian habitat mitigation.

One of the first changes made to the property was its name, from Amen to Kachituli Oxbow. The oxbow was added to describe the geomorphic characteristic intended for the mitigation—a form that was created, not restored. The Sacramento River is precluded from entering the constructed oxbow except under extreme hydrologic conditions. Still, the design was appropriate for the geomorphic setting and allowed for the development of riparian habitat fitting the mitigation objectives.

The topography, soils, and hydrology of the Kachituli Oxbow are similar to those of the project site. The land was extremely flat varying from 12 fmsl to 14 fmsl. The surface soils were Sycamore silty clay loam and Sycamore complex, drained (Fig. 9-7). These soil mapping units were those contained within the landward side of the levee at the Lighthouse Marina project site. The sources of water are precipitation, groundwater, and irrigation. The stratified soil layers, ranging from fine sands to clay, allow some lateral movement of groundwater. Soils of the riverside portion of the mitigation project (outside of the levee) are Valdez sandy loams: these soils are frequently flooded and support good riparian vegetation.

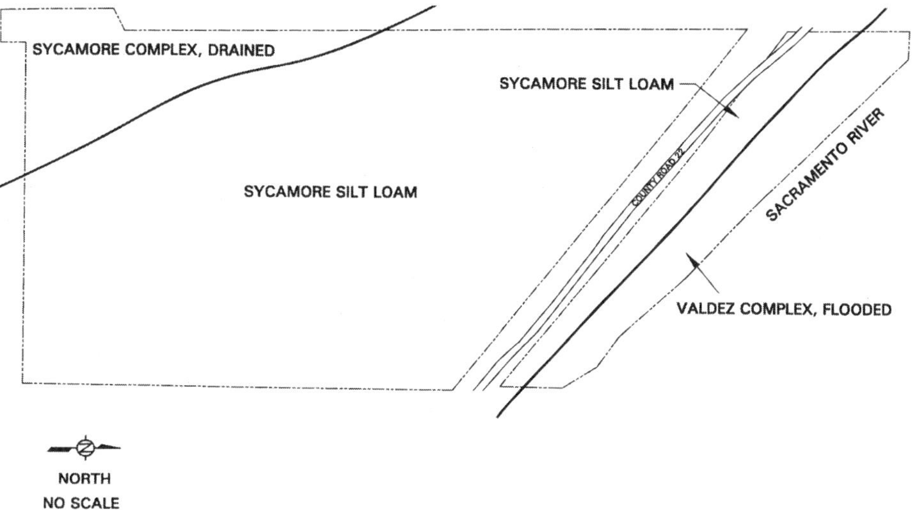

Figure 9-7 Soils of the mitigation site.

The plant community of the mitigation site was considerably different from that of the Lighthouse Marina project site. The Kachituli Oxbow site had been occupied by Europeans since the early 1850s and intensively farmed. In the early years, wheat and oats were grown, but in more recent times tomatoes were the main agricultural crop. A small walnut orchard remains on the site today.

No prehistoric artifacts were known to have existed or were found on the property (Cultural Resources Unlimited, 1990). Still, several agricultural buildings of more recent vintage remained. Most of these were removed with the exception of one large barn that was preserved for storage and may be considered for educational purposes in the future.

The design of the mitigation effort was divided into three phases: topographical, hydrologic, and botanical. The topographical characteristic of the mitigation site was to be that of a meander scar of the Sacramento River (Fig. 9-8). An assessment of soil revealed no old meander trenches. Several oxbow lakes near the site were examined and used as design reference sites. The oxbow meander was designed to have a bottom elevation of 5 fmsl, which is approximately 6 feet above the invert elevation of the Sacramento River adjacent to the property but 3 feet below the normal water elevation. The side slopes of the constructed meander in some cases approached 4 to 1 on the outside edge of the meander but were much more shallow, 20 to 1, on the inside edge, as would have occurred naturally along the river. To create this geomorphic form, more than 250,000 cubic yards of earth had to be moved.

In this low and seasonally distributed precipitation region, given the absence of a surface water connection to the river, the hydrology for the mitigation site

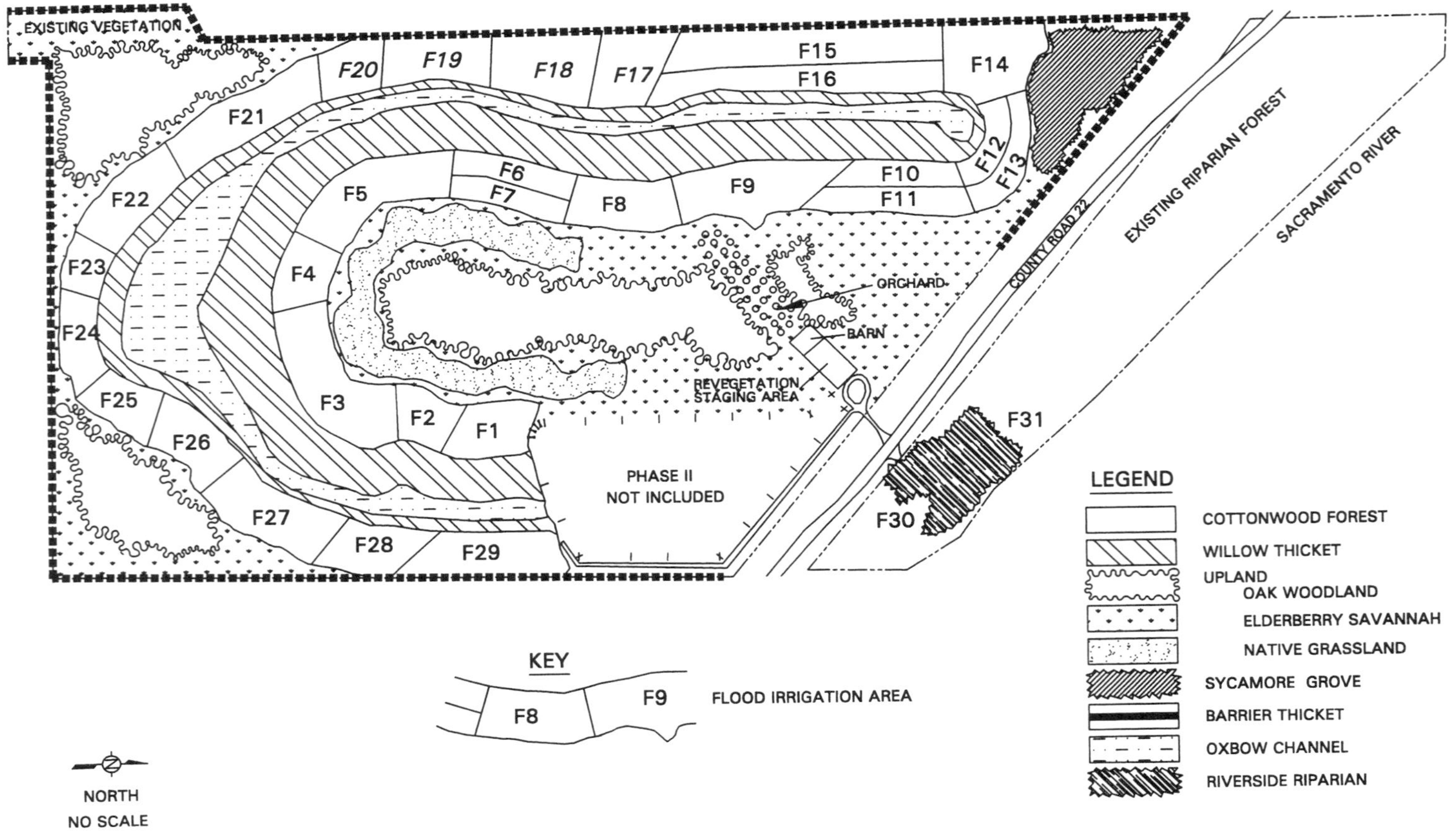

Figure 9-8 Mitigation plan.

needed careful consideration. First, the depth to seasonal groundwater was assessed across the site and the excavation specifications were developed to take advantage of those depths. Then, to ensure the establishment of the intended plant communities, an irrigation system was devised. This system operated with existing and new pumps and water rights associated with the former agricultural uses. However, the design called for irrigation to be terminated after the woody vegetation was established, that is, after their roots had reached adequate groundwater. This addressed one of the design objectives, which was to accomplish a self-sustaining landscape—of course, the landscape might have become more quickly naturalized had it been possible to connect the meander to the Sacramento River. This action, however, would have required that the levee along the Sacramento River be breached, which would have been unacceptable to the many property owners protected by the levee and to the Corps, the agency responsible for its construction and maintenance.

Given the low gradient of the Sacramento River and the large flows that naturally move through the channel system in the spring, the river is flanked by extensive floodplains, at times up to 15 miles wide (Kelley and Green, 1990). In presettlement times, the floodplains were inundated frequently enough to support extensive marshes. Upstream of the Sacramento–San Joaquin Delta, the Sacramento Valley itself contained 550,000 acres of tule marshland (Kelley and Green, 1990), approximately 10% of the watershed area. Adjacent to the river, the natural levees supported lush riparian forests whereas behind the levees, in channels cut off by previous flood events, large shallow oxbow lakes existed. These features set the geomorphic criteria for the design of the mitigation site.

Field surveys were conducted on six oxbow lakes along the Sacramento and Feather Rivers to gather the characteristics for establishing the design criteria. The following information was collected:

- Depth, length, width, and bank slopes of the river channel in the meander area.
- Presence of water on the channel floor.
- General condition of the oxbow with regard to clearing and/or surrounding land uses and access routes.
- Human perceptions and uses of the oxbow areas.
- Soils.
- Plant species and communities.
- Ecological structure.
- Hydrology.

This information served the designers well.

A sinuous meander channel was specified for the site with a deep pool of water, approximately 5 feet, at the southern end. This formed the deepest part of the excavation, which measured 14 feet from the existing land surface. The design was supported by groundwater observations, which were made prior to plan development—groundwater was found to reach within 9.5 feet of the existing soil surface (Kelley and Green, 1990). During excavation, the topsoil was to be saved and used for dressing the finished surface.

Water was intended to accumulate in the "low flow" channel and southern pool of the oxbow. These areas would provide habitat for emergent vegetation such as bulrush and cattail, and fish and waterfowl. On the outside curve of the oxbow, where the bank slopes were steeper, willows were to be planted on the toe of the cut and box elder, buttonbush, and ash were to be planted on the bank slopes themselves (Fig. 9-9). A cottonwood forest was designed for a 150-foot strip at the top of the bank. Elderberry, valley oak, and box elder were to be interspersed among the cottonwoods. Further away from the oxbow, the design called for mixed elderberry savanna and oak woodland. A 2-acre sycamore grove was planned for the northwest corner of the site (Fig. 9-8).

The inside portion of the oxbow was to be planted with willow at the toe, grading to ash and box elder. At the top of the bank a 150-foot zone of cottonwoods would be created. Adjacent to the cottonwood forest, a valley oak woodland and an elderberry savanna would be planted in a strip 145 feet wide and 1,400 feet long.

The design even called for a point bar to be established on the inside of the oxbow. This bar was to have a very gentle slope, 40 to 1, providing habitat for such emergent vegetation as tule, rush, and sedge, as well as horsetail and various species of willows.

The perimeter of the mitigation site was to be screened or buffered by a living fence consisting of blackberry and wild rose. Also, poison oak was con-

Figure 9-9 Wider portion of the oxbow during the dry season with high water indicated by dried algae on shore vegetation. Photograph by David Kelley. Photo also appears in color insert.

sidered for planting in subsequent years if it didn't establish on its own. This fence was intended to discourage people from walking onto the site from the surrounding agricultural areas. No public access was intended for the 5 years during which management and monitoring of the site was to take place, as required by the Corps permit.

Some of the existing plant communities would be left intact. For example, the domestic walnut grove was to be left to provide suitable perches, food, cover, and nesting habitat for a variety of birds, which would be attracted to the site. Also, the plan noted that native walnuts were interspersed throughout the riparian zone along the Sacramento River. Other native vegetation such as valley oak was to be flagged and saved if possible.

The plant communities and specifications were as follows (Miriam Green Associates, 1990):

> Oak Woodland—Approximately 14 acres of oak woodland are designated for the land side of the levee. This area will be planted predominantly with valley oak seedlings and acorns. We will not plant pure stands of oaks; rather, oaks will serve as the dominant species with other vegetation, including elderberry and California buckeye, integrated with the oaks, especially at the edges. Oaks will be planted on 40-foot centers resulting in 25 trees per acre (25 trees per acre × 14 acres = 350 trees). Where direct seeding is employed, three acorns will be placed within each hole. In cases where more than one seedling becomes established, the most robust seedling will be chosen and the others will be removed. Elderberry and California buckeye (*Aesculus californica*) will be planted as secondary species in the oak woodland at a density of 4 trees of each species per acre resulting in 112 trees (56 of each elderberry and buckeye).
>
> Elderberry Savanna—One of the conditions of Permit No. 9051 is the incorporation of a minimum of 500 elderberry plants into the project design to compensate for the loss of endangered species habitat at the Lighthouse Marina project site. The required number of 500 will be planted in an elderberry savanna; 921 elderberries have been incorporated into the entire project. Elderberry will function as the dominant species on 21 acres of elderberry savanna habitat at a density of 25 plants per acres on 40-foot centers, resulting in 525 elderberry plants in this zone (25 trees per acre × 21 acres = 525 plants). Valley oak will be planted in the savanna at a density of 5 trees per acre resulting in 105 trees in 21 acres of habitat. California buckeye will be planted at a density of 4 trees per acre resulting in a total of 84 trees.
>
> Cottonwood Forest—The plan includes 40 acres of cottonwood forest on the land side of the levee. Cottonwoods will function as the dominant species, planted on 20-foot centers, resulting in 108 trees per acre. A total of 4,320 cottonwood cuttings will be planted (108 trees per acre × 40 acres = 4,320 trees).
>
> Valley oak, elderberry, box elder, and Oregon ash will be planted as secondary species at various spacings in the cottonwood forest zone at densities of 2, 6, 8, and 8 per acre, respectively. This will result in a collective total of 960 trees in 40 acres (80, 240, 320, and 320 of each species, respectively). Understory shrubs and fast-growing vines, such as poison oak and wild grape, will not be planted

during the first year. We propose to let these species colonize the site naturally. Wild grape and poison oak are already colonizing the land side where it has been left uncultivated. By not planting shrubs and vines in the first year it will facilitate easier maintenance and will reduce competition with the desired tree species.

Willow Thicket—Our plan specifies 15 acres of willow thicket. Four species of willows (red willow [*Salix laevigata*], sandbar willow [*S. hindsiana*], Goodding's willow [*Salix gooddingi* var. *variabilis*], and Arroyo willow [*S. lasiolepis*]) will be planted as the primary vegetation type. Willows will be planted on 10-foot centers resulting in 435 cuttings per acre. A total of 6,525 cuttings will be required (435 trees per acre × 15 acres = 6,525 trees). Box elder, Oregon ash, black walnut, and buttonbush will be planted as secondary species at densities of 4, 4, 9, and 30 trees, respectively, resulting in a collective total of 705 trees of these four species in the willow thicket.

Sycamore Grove—Two acres are designated as a sycamore grove on the land side of the levee proximate to the railroad tracks. This location was chosen because of the presence of good drainage characteristics of the soil and its suitability for sycamores. Several of the riparian areas we studied, as well as other sites observed by the authors, have remnant stands of the California sycamore. Sycamores formerly were found scattered in riparian forests along the Sacramento River (Thompson, 1961; Holstein, 1984), but have become scarce except for disjunct populations somewhat removed from the densest riparian vegetation. Sycamore seedlings are being contract grown in a nursery and will serve as the nucleus of the population. Sycamores will be planted on 35-foot centers resulting in 54 trees per acre. The 2-acre grove will support 108 trees at this density (54 trees per acre × 2 acres = 108 trees). Valley oak will be planted on the edges of the sycamore grove on 50-foot centers, resulting in 34 oaks scattered along the edges of the grove.

Emergent Vegetation—Cattail, bulrush, and tule rhizomes will be planted in the oxbow in the low flow channel and in the deep water pool on approximately 50-foot centers to promote the establishment of emergent vegetation in the wettest areas. Rootballs approximately 1 foot × 1 foot, including mud substrate, will be planted into holes about the same size. Emergent vegetation is currently scattered along the irrigation ditches; once agricultural practices (including the burning of vegetation along the channels) are discontinued, emergent species are expected to recolonize the bare areas.

Barrier Thicket—The barrier thicket or "living fence" will cover about 4.5 acres around the perimeter of the mitigation site. Initially, we will plant only blackberry and wild rose to establish the fence; other shrubs such as poison oak, coyote brush (*Baccharis pilularis*), and wild grape are expected to colonize this area later due to their presence in several locations on-site and nearby. Clumps of blackberry cuttings will be planted on 15-foot centers along V-ditches resulting in an approximate total of 6,550 blackberry cuttings required for the living fence habitat zone. Clusters of nursery-grown wild rose seedlings will be planted on 5-foot centers, interspersed with the blackberries, resulting in 2,932 wild rose plants in this perimeter zone.

The success criteria for the Kachituli Oxbow mitigation project were quite simple. Based on planting densities for each habitat type (Table 9-1), the gen-

TABLE 9-1 Cumulative Plant Totals

Habitat Type and Species	Acreage	Density (Plants/acre)	Approximate Spacing	Total No. of Plants	Type of Installation[a]
Oak woodland	14				
Valley oak		24	40′ on center (oc)	350	C, S
Elderberry		4	Varies	56	C
California buckeye		4	Varies single spec.	56	S
Sycamore grove	2				
California sycamore		54	35	108	C
Valley oak		17	50	34	C, S
Elderberry savanna	21				
Elderberry		25	40	525	C
Valley oak		5	Varies	105	C, S
California buckeye		4	Varies	84	S
Cottonwood/oak riparian	40				
Cottonwood		108	20’	4,320	P
Valley oak		2	Varies	80	C
Box elder		8	Varies	320	C
Elderberry		6	Varies	240	C
Oregon ash		8	Varies	320	C
Riverside riparian	10				
Cottonwood		54	35	540	X
Valley oak		2	100	20	S
Elderberry		10	Varies	100	C
Box elder		8	Varies	80	C
Oregon ash		8	Varies	80	C
California sycamore (1 acre only)		54	35	54	C
Buttonbush		4	Varies	40	C
Willow thicket	15				
Willows (various spacing)		435	10	6,525	P, X
Buttonbush		30	Varies	450	C
Box elder		4	Varies	60	C
Oregon ash		4	Varies	60	C
California black walnut		9	Varies	135	S
Barrier thickets	4.5				
Blackberry		Varies	15′ on in v ditch row	6,550	X
Wild rose		Varies	5′ oc in v ditch row	2,932	C
Existing wetland area	1.2				
Mud flat	3.5				
Project Totals	111.2	Varies	Varies	24,224	

Source: Miriam Green Associates, 1990.
[a]Key: C = container, S = direct seeded, P = pole cutting, X = wattle or cutting.

eral criterion was 80% survival of each tree community. This criterion was then applied to each tree species (Table 9-2). No specific criteria were established for the presence of understory cover or non-native vegetation or for the percent cover achieved by the various installed plants and structured plant communities.

TABLE 9-2 Five-year Success Criteria

Species	Initial Planting	20% Loss	80% Survival
Box elder	140	92	368
California buckeye	140	28	112
Buttonbush	490	98	392
Oregon ash	460	92	368
Black walnut	135	27	108
California sycamore	162	32	130
Cottonwood	4,860	972	3,888
Valley oak	589	118	471
Wild rose	2.932	586	2,346
Blackberry	6,550	1,310	5,240
Elderberry	921	184	737
Willow	6,525	1,305	5,220
Total no. plants	24,224	4,845	19,379

Source: Miriam Green Associates, 1990.

The success criteria made no reference to the presence or absence of wildlife, but wildlife was included in the monitoring plan.

A 5-year monitoring program was required by the Corps and specified in the special conditions attached to the permit. Aerial photographs were to be taken to document the changes in canopy cover. Trees were monitored for survival and growth. Aerial photographs taken in 1990, at the outset of the restoration process, were to serve as a baseline for judging progress. At the end of the third year, the monitoring plan called for the regulatory agencies to review progress. This review was intended to permit adjustments to the management program to ensure the achievement of the 5-year goal. Both the planted species and those invading the site would be monitored in order to assess the maturation of the various plant communities. Although not identified as such, the review would meet the principle of adaptive management.

Management of the restoration process was primarily focused on irrigation of the newly installed plants. Water rights were associated with the property and these rights were to be exercised in providing adequate moisture for the establishment of the various plant communities ranging from those adapted to open water to those of upland character. Three methods of irrigation were employed: flood, sprinkler, and drip. It was anticipated that 3 years of irrigation would be necessary to establish the various plant communities. At the end of 3 years, if irrigation were still needed, it would be continued.

Flood irrigation was to be used in areas having reasonably flat slopes, between 0 and 2%. Using the existing pumping station on the Sacramento River, water would be lifted to trenches that would then distribute the water throughout the designated irrigation area. Water would generally be pumped during the dry season, from April through October.

Willow thickets would be irrigated with sprinklers. Again, it was anticipated that irrigation would be needed for the first 2 to 3 years, until the roots of these plants developed. Drip irrigation was intended for use in the oak woodland, elderberry savanna, and sycamore grove plantings. Water would be applied to these areas from May through September.

Weed control was given some emphasis in the management plan. Because of the surrounding agricultural activities, a number of weed species potentially could invade the site. The weed control techniques were to include disking, herbicides, the use of weed abatement fabric, and hand removal. Following the last tomato harvest, the entire site was to be disked.

One final concern of the project designers was that of predation by rodents, particularly mice. Plant collars and protective screens were designed to be placed around seedlings. These collars also were intended to protect roots and help concentrate water into the root zone.

The monitoring program recognized the importance of wildlife to the intended landscape. A wildlife survey methodology was developed including the following categories: birds, mammals, reptiles, and amphibians. Neither the anticipated species nor the sampling methodology were specified.

Finally, hydrologic monitoring was specified. This involved the installation of five water stage gages capable of measuring to the closest 0.1 foot. These gages were distributed along the oxbow. No groundwater monitor was included at this point, although the deepest part of the oxbow would reflect local groundwater elevations, particularly during the dry months of the year, March through October. The measurements from the stage gages reflected the hydrologic effects of precipitation and local runoff, groundwater movement, and irrigation.

The final monitoring program was published in August, 1991 (Jones & Stokes Associates). The plan was assembled some 2½ years after the permit was issued, and the plan anticipated further delays.

MITIGATION IMPLEMENTATION AND RESULTS

In September, 1989, the Corps allowed the proponents of the Lighthouse Marina project to proceed with clearing and grading activities on the project site. The final restoration plan was approved by the regulating agencies in October, 1990. Shortly thereafter, excavation of the oxbow began. Following excavation, planting was started in July, 1991. Because parts of the irrigation scheme did not work properly, valley oak acorns were planted during the fall of 1991 at higher densities than specified.

At the end of the first year, 1992, a healthy cover crop of clover prevented weedy species from invading the site. Some desirable plant species such as sandbar willow were beginning to self-propagate. The majority of the plant materials survived, and the barrier thicket surrounding the restoration site prospered during the first year. In the riverside riparian areas, more than 95% of the plant material survived and was growing vigorously. In this area, weed propagation, at the end of the first year, did not seem to be a problem.

The dense weedy ground cover in the valley oak woodland made it very difficult to locate the oak seedlings. Based on those located, approximately 25% of the seeds had germinated and survived the first year, 1992. All in all, the first-year progress in establishing the intended plant communities seemed to be successful. Several recommendations, however, were forthcoming: manual or chemical removal of weeds around cottonwood and willow trees, removal of cover crop duff around the perimeter of the trees, periodic installation of replacement plants in the upland monitoring units to compensate for additional mortality, installation of raptor perches to encourage rodent predation (rodents were indeed destroying some of the introduced woody species), and irrigation of the valley oak seedlings during the dry season.

Only birds were monitored, or at least noted, in the first year's monitoring report (Jones & Stokes Associates, 1992), which was submitted to the Corps. A total of 43 bird species was observed during the winter and spring surveys. The bird populations were divided among the various landscape units. Eleven species were observed in the upland units, six species in the oxbow lake and nine species in the woodland habitats. A similar distribution was observed during the spring survey. A few mammals, or their signs, were observed: black-tailed hares, raccoons, and California blacktail deer.

At the end of the second monitoring year, 1993, the problem of rodent herbivory and invasion of exotic grasses had become a major problem and required immediate action. Management procedures were set out for subsequent years. During the second year, bird usage of the site increased dramatically in terms of species. Fifty-nine species were observed in the second survey (Jones & Stokes Associates, 1993).

By 1994, the plant communities had matured and had exceeded the performance anticipated for this stage of development (Jones & Stokes Associates, 1994). Rodent herbivory was controlled, but the invasive exotic grasses and weeds required continued monitoring and management. The more mature state of the cottonwoods, willows, elderberries, and grasslands provided better habitat values and consequently continued to retain and attract bird populations. The surveys associated with the third-year monitoring report show that the bird species had stabilized—56 species were observed in the 1993–1994 survey as opposed to 59 species observed in 1992–1993.

By the fourth year of monitoring, 1995, the plant communities seemed to be on track to meet the 5-year performance criteria. Despite a problem with the sycamore trees, the survival and growth rates of the woody vegetation were reported to be satisfactory. Only minor adjustments to the weeding and irrigation programs were recommended (Jones & Stokes Associates, 1995).

The number of bird species using the oxbow portion of the site fell during this monitoring period; however, this was considered to be a temporary phenomenon, largely attributed to the prolonged wet and cold spring and inunda-

Figure 9-10 Kachituli Oxbow on west bank of the leveed Sacramento River. Photograph by David Kelley. Photo also appears in color insert.

tion of the mud flats that had developed around the open water areas. The number of species observed over the entire site fell from 56 to 34. Still, the well-established plant communities and their usage by wildlife convinced the Corps that a fifth year of monitoring was not necessary. On January 27, 1997, the Corps concurred with the designers and implementers of the Kachituli Oxbow mitigation project that the requirements of the 404 permit had been met. The Corps called the mitigation "successful and complete." They concluded that the Kachituli Oxbow was in compliance with the terms and conditions of permit 9051 (Fig. 9-10).

CONCLUSIONS

The success of the project stems directly from the interest and involvement of regulators and citizen groups and the expertise of the designers. On first analysis, the concept for the Kachituli Oxbow mitigation plan was rather radical —converting a tomato field to riparian habitat. Although in geologic time there may have been an oxbow present on the property, there certainly was no evidence of this geomorphic structure on the preexisting alluvial soils. However, careful analysis of the soils and soil stratigraphy in advance yielded a plan and project that were successful. Specifying the geomorphic form, of course, was only part of the conditions for success. There were two remaining components: the hydrology and plant communities.

Predesign analysis of the site revealed the presence of groundwater, which would be intercepted by the intended geomorphic structure. This knowledge ensured the presence of an adequate, sustainable water supply. During the transition and establishment of plant communities, this supply had to be supplemented, as the designers clearly understood. Consequently, an elaborate irrigation system was put in place and used to provide the necessary moisture for plant establishment.

The plant communities and their appropriate landscape positions were well thought out and implemented. This was clearly facilitated by the predesign analysis of similar habitats that still existed along the Sacramento River. The firsthand experience and understanding of these habitats contributed significantly to the success of the Kachituli Oxbow mitigation project. Another contributing factor was the ongoing management and monitoring, which is to say adaptive management. When weeds and herbivory became a problem, they were overcome by aggressive action, which prevented the mitigation from taking an unwanted course.

The FWS also should be given credit for the success of the project. Although they did not have a direct hand in the design or implementation of the mitigation project, they helped steer and focus the effort. Through a number of years of contentious negotiations with the project proponents, they maintained their focus on habitat preservation and restoration. Recognizing that the project proponents would ultimately be given permission to proceed with construction, the FWS and the numerous environmental groups concentrated on how best to make the mitigation successful. In the end, the evidence indicates that this was accomplished and that the environment, wildlife, and citizens of the region benefited.

On August 16, 1997, the title to the Kachituli Oxbow property was transferred to the California State Lands Commission. This agency will be responsible for the long-term management of the property. The intention is to open the project to public use with an emphasis on education.

ACKNOWLEDGMENTS

Representatives of the regulatory and resource agencies, design firms, and interested parties participated in the preparation of this case study. Although some were not involved in the early stages of the project's development, they were all well informed on the history, objectives, and technical issues of the project. June de Weese with the FWS in Sacramento, California, spent a great deal of time discussing the project with the authors and providing background information. She coordinated a site visit, which included representatives of the Corps, the EPA, and the California State Lands Commission. Also in attendance were the designers and implementers of the project including Miriam Green of Miriam Green Associates, David Kelley of Kelley and Associates Environmental Sciences, Inc., and Ellen Davis of Jones & Stokes Associates, Inc. A number of people have written about the project and their documents were used extensively. Where particular concepts or observations were noted, every attempt was made to credit the author properly.

REFERENCES

Andrews, W. F., *Soil Survey of Yolo County, California*, U.S. Government Printing Office, Washington, DC, 1972.

Corollo, P. S., *Letter to District Engineer, Sacramento District, Army Corps of Engineers*, Lighthouse Marina and County Club, West Sacramento, CA, 1990.

Cowardin, L. M., V. Carter, F. C. Golet, and E. T. LaRoe, *Classification of Wetlands and Deepwater Habitats of the United States*, U.S. Fish and Wildlife Service, FWS/OBS-79/31, Washington, DC, 1979.

Cultural Resources Unlimited, *A Cultural Resources Study for Kachituli, Yolo County, California*, Lighthouse Marina and Riverbend Development, West Sacramento, CA, 1990.

Dahl, T. E., *Wetlands Losses in the United States 1780s to 1980s*, U.S. Department of the Interior, Fish and Wildlife Service, Washington, DC, 1990.

ECOS, *Site Assessment for the Amen Property*, Lighthouse Marina and Riverbend Development, West Sacramento, CA, 1989.

EDAW, Inc., *Draft Environmental Impact Report and Environmental Impact Statement for the Lighthouse Marina*, Yolo County, Community Development Agency, Planning Division and U.S. Army Corps of Engineers, Sacramento, CA, 1986.

EDAW, Inc., *Final Environmental Impact Report and Environmental Impact Statement for the Lighthouse Marina, Broderick, California*, Yolo County, Community Development Agency, Planning Division and U.S. Army Corps of Engineers, Sacramento, CA, 1986.

Green, S., "State Doesn't Want Land Sought for Broderick Marina," *The Sacramento Bee*, Sacramento, CA, 1985.

Hansen, H., *California, A Guide to the Golden State*, Hastings House, New York, 1967.

Holstein, G., "California Riparian Forests: Deciduous Islands in an Evergreen Sea," in R. E. Warner and K. M. Hendrix, editors, *California Riparian Systems*, University of California Press, Berkeley, California, 1984.

Jones & Stokes Associates, Inc., *Kachituli Oxbow Mitigation: First-Year Monitoring Report*, Lighthouse Marina and Riverbend Development, West Sacramento, CA, 1992.

Jones & Stokes Associates, Inc., *Kachituli Oxbow Mitigation: Fourth-Year Monitoring Report*, Lighthouse Marina and Riverbend Development, West Sacramento, CA, 1995.

Jones & Stokes Associates, Inc., *Kachituli Oxbow Mitigation Monitoring Program*, Lighthouse Marina and Riverbend Development, West Sacramento, CA, 1991.

Jones & Stokes Associates, Inc., *Kachituli Oxbow Mitigation: Second-Year Monitoring Report*, Lighthouse Marina and Riverbend Development, West Sacramento, CA, 1993.

Jones & Stokes Associates, Inc., *Kachituli Oxbow Mitigation: Third-Year Monitoring Report*, Lighthouse Marina and Riverbend Development, West Sacramento, CA, 1994.

Katibah, E. F., "A Brief History of Riparian Forests in the Central Valley of California," in R. E. Warner and K. M. Hendrix, editors, *California Riparian Systems*, University of California Press, Berkeley, CA, 1984.

Kelley, D. B., and M. Green, *Soils of the Kachituli Oxbow, Yolo County, California*, Lighthouse Marina and Riverbend Development, West Sacramento, CA, 1990.

Kelley, R., *Battling the Inland Sea*, University of California Press, Berkeley, CA, 1989.

Kroeber, A. L., *Handbook of Indians of California*, Smithsonian Institution, Dover Publications, Inc., New York, 1925.

Levin, M., *Letter to Donald Hadel, Secretary of the Interior*, Levin Associates, Carmichael, CA, 1985.

Miriam Green Associates, *Kachituli Oxbow Revegetation Plan*, Lighthouse Marina and Riverbend Development, West Sacramento, CA, 1990.

Miriam Green Associates, *Kachituli Oxbow Technical Specifications*, Lighthouse Marina and Riverbend Development, West Sacramento, CA, 1990.

Spotts, R., *Letter to Tom Coe, Regulatory Section, Sacramento District, Army Corps of Engineers*, Defenders of Wildlife, Sacramento, CA, 1986.

Thompson, K., *Riparian Forests of the Sacramento Valley, California*, Annals of the Association of American Geographers, Cambridge, MA, Volume 5-1, 1961.

Sacramento River Preservation Trust; Davis Audubon Society; Defenders of Wildlife; Mother Lode Chapter, Sierra Club; Sacramento Audubon Society; Planning and Conservation League; American Fisheries Society; California Native Plant Society; and Environmental Council of Sacramento. Letter to Colonel Scholl, U.S. Army Corps of Engineers. Sacramento, CA, 1986.

U.S. Army Corps of Engineers, Sacramento District, *Public Notice and Scoping for Joint EIR/EIS No. 9051*, Sacramento, CA, 1985.

U.S. Army Corps of Engineers, Waterways Experiment Station, *Preliminary Guide to the On-Site Identification and Delineation of the Wetlands of the Interior U.S.*, Technical Report Y-78-6, Vicksburg, MS, 1982.

CHAPTER 10

CONCLUSIONS

The case for restoration will be argued as long as there remains one person to challenge it. Still, the case studies and many other examples show that restoration can be successful. The unique, rare wetland types will always bolster the opposing argument, until enough is known about them and sufficient experience gained with their restoration. Perhaps, for such wetlands, preservation is the only answer. In the meantime there is a great need for wetland restoration in the United States, and sufficient knowledge is available to reclaim successfully a wide variety of wetland types. Not only have we learned how to restore and re-create our wetland resources, but we know today that where we do so we solve important water resources problems and improve the quality of our lives. That we need more wetlands in this country is undeniable. We have already lost more than half our wetlands since presettlement days through the economic activities necessary for creating a modern nation—agricultural, commercial, and industrial enterprise. We will continue to lose small quantities, no doubt, to necessary economic projects, but it is not unreasonable to expect that by the year 2010 we can accomplish the National Research Council goal of an additional 10 million acres of restored or recreated wetlands.

As we have discovered in the past few decades, wetlands make our country a better place to live. Wetlands provide us with more ducks and geese to travel our migratory flyways, less devastating flooding along our major rivers, and clearer and healthier surface waters in which to swim and fish. And who knows how many of the poorly understood problems that loom on the horizon, such as hypoxia in the Gulf of Mexico and sea-level rising from global warming, can perhaps be ameliorated by increases in wetland resources.

The emphasis on wetland restoration need not and must not diminish the importance of protecting our existing wetlands. Avoidance and minimization should continue to play the key role in the Section 404 permitting process. But that same process can allow us to go well beyond the narrowly focused quid pro quo of "take an acre here, restore an acre there." A large-scale increase in wetlands in North America should be our goal.

The science of restoration is in its infancy, yet the many practitioners in the field have made great strides since the early 1970s. Those kinds of restorations that can be done well, should be done; where uncertainties remain we should proceed with caution; and where rare wetland types exist that cannot be duplicated, they should be preserved at all costs. In the meantime, we continue to learn.

The four restoration projects that have been described in the preceding pages are judged to be successful by those most closely involved. Success criteria applied to restoration projects range across a wide variety of scientific standards, from species-specific survival rates to general hydrologic conditions. Such criteria, whether defined by the public or the scientist, are important to establish in order to guide landscape development to the desired goals. Setting national or even regional success criteria will never work. The goals, objectives, and criteria should be established in relation to the water regime of the drainage basin and ecosystem in which restoration is intended. In this context it is possible to address the often neglected wetland benefits of water quality management, flood control, and erosion control, which should be reflected in the success criteria for every restoration project. Even if the specific biological criteria for a mitigation project are not immediately satisfied, these other wetland functions may be exhibited, facilitating biological success at a later date. Wetlands failing to meet specific habitat can still store water, preventing downstream flooding as well as providing adequate detention time for the removal of sediments, nutrients, and other constituents. Although at times the short-term goals, by which the success of the mitigation project was defined, may not be met, the restoration of topography and hydrology, on the land that has been set aside, affords future opportunity for meeting them.

Successful restoration projects will be found appropriately located in the landscape. An inappropriately placed wetland may negate the objectives and criteria for its restoration or creation. The land benefits can be realized only if they are perceived and utilized by the people that will use them and are affected by them. Wetlands restored in locations remote from public access will have failed to meet their full potential by benefiting wildlife but not satisfying recreational, aesthetic, or educational needs. The landscape position is also important for the establishment of the all-important hydrologic characteristics. Location of restored wetlands too low on the landscape may result in excessive inundation and/or the development of a type of wetland not critically needed. When they are positioned too high on the landscape, on the other hand, the supply of water essential for successful restoration may not be available. The models for successful landscape position and hydrology, as well as for plant communities and other attributes, often exist in natural landscapes adjacent to or near the restoration site. Basing design parameters on the proven relationships within extant landscapes similar to the desired wetland type has been shown to be a successful strategy.

Successful restoration projects will involve a wide variety of interest groups and institutions. The role of the scientist, particularly in the botanical disciplines, is well established. What is less clearly appreciated is the role that the governmental institutions and the public play. The case studies showed the value of a keen interest on the part of the professional staff guiding the regulatory and mitigation processes, as well as the equally keen and perhaps even belligerent involvement of the public. Properly focused, the dialogue between various interests is important to the development of successful mitigation projects. Without the legislated regulatory process, this dialogue would not likely take place, and certain interests would be depreciated and ignored. The NEPA and Section 404 of the Clean Water Act have become essential elements in this dialogue and have promoted effective and beneficial restoration of aquatic landscapes.

Successful restoration projects will foster an active dynamic process that extends from early planning far beyond the completion of construction into management and monitoring. Agreement on mitigation design is only the first step. One of the most important steps, construction, is frequently downplayed. Construction includes grading, planting, and placement of hydraulic control structures. As these elements are put in place in various combinations, the complex construction activities need careful supervision. Mitigation expertise should be applied at every level of construction management and resulting landscape development. Although design drawings and construction specifications are important, a clear understanding of the goals and objectives of restoration and the construction process by the supervising engineer or scientist is essential in accomplishing a successful project and can even compensate for the absence of such documentation. Monitoring is the next step in the process and, without monitoring, effective management cannot take place. Conversely, without management, the monitoring effort is largely wasted. Clearly, short-term monitoring and management are essential to meeting the goals set for the restoration project, and long-term management and monitoring are essential to sustaining the benefits resulting from that restoration.

Successful restoration projects may have led the participants in directions they could have never anticipated. Although most engineers (and even scientists) believe that the application of well-researched principles will lead inevitably to a particular mitigation result, the reality does not support this view. In the case studies and in many other restoration projects across the country, serendipity often plays a significant role and can be used to advantage by creative and flexible practitioners.

Successful restoration projects will have no terminal point. The completion of a restoration project is only the first step, the beginning of the life of the restored landscape. The survival of that landscape and the gathering of benefits from its involvement in the environment will be sustained only if the landscape continues to prosper, and for this, long-term ownership and management is essential.

The four restoration projects described in the preceding chapters provided some insights into the elements that constitute success, leading to the general conclusion that there is no single way to reach that goal. In looking at the planning, the implementation, and the role that people both officially and unofficially played in their development, certain themes emerge from an examination of the four projects. The following observations and recommendations are based on the analysis of the case studies in the context of the subjects discussed in earlier chapters: their role in the natural landscape, the economic importance and impact of agricultural drainage, the origins and substance of the federal legislation that protects and preserves our wetlands, and the key role that compensatory restoration plays in federal wetlands policy today.

PLANNING AND DESIGN

Ideally, the planning and design process for any mitigation project should begin at the watershed level, with an appreciation of the basin-wide effects of wetland losses and wetland gains guiding the design process. Although, for example, flood control and water quality improvement cannot be achieved in a 60,000-acre watershed by the addition of one or two 5-acre wetland mitigation projects, the strategic placement of 100 or 1,000 of those small projects may have a substantial accumulative effect.

Watershed planning for wetland protection and restoration was not in vogue when the projects under discussion were executed but, to the extent that the broader ecological and hydrological regimes were understood and taken into account in their development, these projects benefited. The damaging impacts of the construction projects were understood in the context of regional needs: the riparian and elderberry beetle habitat needs that were satisfied by the Kachituli Oxbow were defined in terms of first watershed and then national deficiencies; the Keys bridges mitigation satisfied the requirements of the broader ecosystem that encompassed the Florida Bay and extended into the Atlantic Ocean; both the Hoosier Creek and Yahara River Marsh projects considered and were sensitive to impacts upstream and downstream from the specific projects.

The consideration of whether the mitigation activities should take place on or off the project site and provide in-kind or out-of-kind restoration varies by individual project. Planners in the case studies made every attempt to place the restorations at the point of impact, with varying success and significance. The Yahara restoration projects were implemented in the complex of marshes being affected, but in Hoosier Creek, the Kachituli Oxbow, and the Keys bridges projects, on-site mitigation proved to be either impossible or far less productive than the off-site restorations that were ultimately accomplished. Restoration of a riparian habitat adjacent to the intense urban landscape along the Sacramento River made far less sense than to create the Oxbow, because the continued disturbance of the former site would have diminished if not destroyed its effectiveness, and the Kachituli site was able to provide the needed habitat and desired benefits far more efficiently. The Hoosier Creek restoration not only produced the desired compensation for the lost wetlands, but was also able to correct severe environmental degradation that occurred in the past. Likewise, by going off site in the Florida Keys, the project designers were able to restore and enhance, with a minimum of effort and investment, hundreds of acres of former wetlands that had been rendered useless by past construction projects. In the latter case, however, the off-site restoration of seagrasses and mangroves did nothing to reduce the erosional impacts that they could have served to ameliorate by stabilizing the soils at the bridge locations themselves.

Some wetland functions, such as groundwater recharge, flood control, and water quality management can be provided by a wide variety of wetland landscapes, which renders moot the tradeoffs between in-kind or out-of-kind mitigation. Yet wetland mitigation is often tied to a particular plant community or habitat. The replacement of shrub carr by sedge meadow, as was the case at Hoosier Creek, resulted in an out-of-kind mitigation. The end result was still highly desirable and acceptable, although the habitats are quite different. A useful exercise would have been to assess the two communities, their roles and potential benefits or deficiencies in the broader watershed and ecosystem. In this context, either habitat may have been acceptable. Further, most of the organisms, including the plants, could move between the site of impact and restoration, via Hoosier Creek or some other vector. Consequently, the tradeoffs between the proximity of mitigation to impact and the exchange of wetland types should be carefully considered in the planning and design process. And, given the unpredictability of the mitigation process, a range of wetland conditions might better be considered rather than the highly specific determination of a single landscape type and specific location. The design might involve alternative habitats that are needed in the region and may well develop despite the designer's best intentions.

The mitigation ratio, that is, the mitigation area divided by the affected area, is increasingly a subject of debate. It is in the interest of the developer, whether public or private, to make the ratio as small as possible. It is in the public interest and therefore that of the regulatory agencies, on the other hand, to increase the ratio as much as possible. These agencies argue that the uncertainty of successful restoration makes it necessary to require larger areas to be restored than were lost to compensate for those mitigation projects that fail, and there is every indication of a significant failure rate. The common practice today is to require a minimum mitigation ratio of 1.5:1 or larger. In some cases, this ratio includes buffer surrounding the wetland or uplands within a matrix of wetlands to provide complex edge and diversity of habitat. In the case of the Yahara River Marsh, the ratio was 1:1, but this was a very early mitigation project. In the case of the Kachituli Oxbow, the mitigation ratio was as high as 2.5:1 for specific habitat types. In Hoosier Creek the ratio was 1:1, but was larger in the end, and in the Florida Keys, what started out as 1:1 ended up vastly greater.

The planners and designers for each of the case studies established success criteria. These criteria, in every case, focused primarily upon the plants and plant communities. By association, wildlife habitat structure was inferred, but no specific analyses or design parameters were established for this habitat except in the case of the elderberry beetle (not a wetland habitat) at the Kachituli Oxbow. Even at the Florida Keys project, where hydrology and water quality issues were important to the planners, the success criteria were limited to numbers of mangroves and acres of seagrass planted and surviving. None of the mitigation project designs or evaluations specifically addressed flood storage, water quality, recreation, or aesthetics. In the designs for the Yahara River Marshes, the riparian habitat associated with the Kachituli Oxbow mitigation project, and the inland lagoons revitalized in the Florida Keys, there were no specific plans developed for public access or use of the mitigation sites. Yet aesthetics were no doubt considered unofficially by the designers in selecting the plants and structuring the plant communities.

As with aesthetics, landscape position was a design consideration that was not explicitly addressed yet was incorporated in each project. The existing matrix of marshes in which the Yahara River mitigation project took place defined and fixed the landscape position of each of the individual restoration projects. In the case of Hoosier Creek, the design engineers established an appropriate hydrologic position in the landscape by installing the downstream hydraulic control structure, improved by the additional adjustments made by beaver, shortly thereafter. The end result was a landscape properly positioned. The inland lagoons where tidal reconnections were established, in the Florida Keys, were compared to and replicated the naturally existing lagoons.

One very important design activity that was illustrated in the Kachituli Oxbow project was direct emulation of the landscape. In this case, the designers went into the field and measured the prototype model, which was riparian habitat. They selected existing riparian habitats close to the mitigation site and then developed plant lists, tree species densities, hydrologic settings, and a variety of other parameters. They then used these parameters to design a riparian habitat associated with the sculpted oxbow. In reality, the designers of the other mitigation projects engaged in a similar process but not so specifically. The surrounding marsh along the Yahara River provided the designers of this mitigation project with firsthand observations of the plant communities and landscape positions, as did portions of the upper Hoosier Creek environment. The difficulties in establishing new seagrass meadows along the Florida Keys, in retrospect, could have been reduced if the emphasis had been put on duplicating the basic conditions under which seagrasses thrive. The mitigation effort that relied upon planting seagrass sprigs near the bridges was largely unsuccessful; the conditions that were created in the lagoons by reestablishing tidal connections, on the other hand, produced the spontaneous development of new seagrass beds.

In all four case studies, hydrology was the major design consideration. Groundwater observations were made at Yahara River, Hoosier Creek, and the Kachituli Oxbow. Ocean tidal flows were a critical design factor in the case of the Keys bridges. Surface water was the dominant force for Hoosier Creek and the Yahara River Marshes, inundating both sites and controlling groundwater elevations. Little or no surface water influences existed at the Kachituli Oxbow, this system being driven almost entirely by groundwater and local precipitation. If the lack of proper understanding of hydrology is one of the major contributing factors to failed wetland mitigation projects, the four case studies illustrate the benefits of understanding the hydrologic regime associated with restoration.

The four case studies illustrate very effectively the design considerations of successful mitigation. None of them, on the other hand, dealt explicitly with the basin-wide considerations of flood control, water quality management, or recreation, nor was consideration given to the delivery of benefits to potential consumers.

IMPLEMENTATION

Implementation of wetland mitigation consists, in each of the four case studies, of construction, management, and monitoring. Construction activities involve clearing, grading, planting, erecting control structures, and the very important management of these activities. Grading may take many different forms and be utilized to accomplish a number of objectives. At smaller restoration projects, such as Hoosier Creek, a backhoe is of sufficient capacity to remove unwanted materials and level the restoration site. In other cases, as in the Kachituli Oxbow project, mass grading is done by bulldozers and scrapers. Trucks may be necessary to remove the unwanted material, such as the foundry sand removed from the Yahara River mitigation project, and again, backhoes may be used to help place the hydraulic control structures, such as the one used in Hoosier Creek, to raise the water levels upstream of the railroad embankment, and to install the culverts that allowed the tidal flushing in the Florida Keys lagoons. Planting is done in a variety of ways, using a variety of materials. Without the necessary expertise at hand, grades can be missed, control structures inappropriately located, and plantings done incorrectly. Problems with materials or procedures arise at every step. In all the case studies, a number of people participated in construction supervision. Despite careful observation, errors can and do occur. The wrong species of plants were installed in portions of the Yahara River Marsh, later detected by the construction managers, and replaced.

In the four case studies independent construction managers or supervisors, representing the mitigation interests, were present. These managers were independent of the construction contractors, which gave the observers the degree of independence and freedom necessary to ensure quality work. It also made the necessary expertise available to deal with the inevitable design changes which occur in the field. Trained scientists and engineers such as those working on the Yahara River and Kachituli Oxbow projects were able to maintain flexibility and, most importantly, they were willing to experiment with the changes during the course of construction. The Florida Keys project, particularly, was implemented in the early days of wetland mitigation, and little experience was available to guide the construction of mitigation. Without fixed rules and long tradition, there was a greater need for advice and guidance to be provided by experienced field observers.

Management and monitoring are an essential part of the mitigation process, and construction should not be considered complete until the objectives have been confirmed by monitoring. The extent and appropriateness of the management programs depended on gathering information about the establishment of the appropriate hydrology and the development of the plant communities. Failures in mangrove propagation led to the search for new locations in the Florida Keys; rodent and invasion of exotic plants led to new management decisions in the Kachituli Oxbow; the implementers at Hoosier Creek had to react to beaver activity; and the problems with Yahara River Marsh plantings required last minute changes. Without this kind of monitoring information, the management of these projects would be a wild guessing game likely to lead to the loss of many of the desired and intended attributes of the new landscapes. The monitoring data alerted the managers to the necessity of providing irrigation water, slightly altering grades, and combating predation and weed infestations. This information made possible the survival of seedlings in the riparian habitat of the Kachituli Oxbow and the ultimate development of sedge meadows in the Yahara River Marsh.

The extent of the monitoring phase varied among the case studies. A minimum of 3 years was required for all and, in some cases, as many as 5 years. At the end of the monitoring program for three of the projects, the regulatory agencies were unanimous in their agreement of the success of the new landscape. In the Yahara River Marsh and at Hoosier Creek, however, the implementing agencies continued to monitor the sites. In the fourth, in the Florida Keys, the regulatory agencies essentially gave up on the seagrass mitigation, only to be rewarded 10 years later by evidence of successful restoration of beds as a result of the project. This was the only example of true long-term monitoring, but it was only in the case of the Keys project that sufficient time had elapsed to allow it. Long-term monitoring, perhaps on a periodic basis, could have considerable value in the assessment of restoration techniques, the evaluation of the sustainability of mitigation wetlands, and the determination of the landscape value to the surrounding community. Synoptic monitoring every 5 years may be sufficient to provide this information.

Despite all the careful planning and design of the mitigation projects, an element of serendipity was important in two of them. The Hoosier Creek mitigation project was intended to produce a shrub carr habitat, but ended by providing a good measure of sedge meadow. Unanticipated by the designers, the remnant sedge meadow began to assert its landscape position and extend into the restored area, replacing the willows that had been consumed by elk and other herbivores. The landscape that resulted from this chance occurrence was as much admired and highly valued as the intended shrub carr could have been. Similarly, the failure of the initial mitigation for the lost seagrass beds along the route of the Florida Keys bridges was more than compensated for in the lagoons that were flushed for the purpose of stimulating mangrove restoration. These accidental results added substantially to the overall value of the restoration projects.

INSTITUTIONAL AND PUBLIC COMMITMENT

Without the regulatory requirements and supporting guidelines of Section 404 of the 1972 FWPCA, along with the environmental reviews required by the NEPA of 1969, none of the ample benefits of the case studies would have been achieved. The conditions under which the bridges carrying the Flagler Railway were first constructed, which must have caused substantial environmental damage, would never be permitted today. These two pieces of federal legislation, alone, have brought a variety of public and private individuals and institutions into the process that provide both the expertise and the interests that produce environmentally sound development projects. They have established the priority for wetland protection and mitigation and set the forum for public debate.

Even in the early stages of their influence, these pieces of federal legislation forced the various development organizations not only to mitigate for wetland losses but also to avoid and minimize substantial environmental damages in the design of the development projects themselves. The Yahara River Marsh project, first proposed in the 1960s and finally completed in 1988, was the first wetland mitigation project undertaken by the Wisconsin Department of Transportation. During the long and drawn out negotiations with the environmental interests, design changes were made that reduced wetland acreage to be destroyed from the original 72 to 31 and finally down to 22 acres, and actual construction resulted in a loss of only 18.3 acres. The Florida Keys project was initiated before the Corps had established procedures for implementing the 404 permitting process. Yet by forcing all the parties to sit down at the same table, with the threat of permit denials driving them to negotiate, substantial accommodation of environmental interests was achieved before the actual permits were applied for and engineers were prevented from replacing the old bridges with causeways that would have affected the complex hydrology of the region. In the Kachituli Oxbow project, the applicants reduced the acreage of destruction of elderberry beetle and riparian habitat before construction occurred.

The process, in all the cases, produced vigorous public debate about the pros and cons of the project and the associated mitigation needs. Both in Wisconsin and in California, these debates elucidated the importance of the wetlands or habitats being lost and the specific needs for their restoration. Despite the antagonism between proponents and opponents of the project, in the end they worked together to produce mitigation successes. The vigorous debate resulted in a new understanding of the landscape functions to be restored, the quality of restoration necessary, and even the techniques to be employed.

Although the end result ultimately benefited the environment, the debates over the project were not always pursued for the sole purpose of preserving the environment. The opponents of the new road alignment across the Yahara River and the replacement of the Florida Keys bridges were as much concerned about encroaching urban sprawl as they were about environmental protection. Similarly, the public uproar that produced the excellent Kachituli Oxbow wetland restoration was really about the elderberry beetle.

For whatever reasons, a cadre of people began to form in all the case studies, in both the public and private sectors, that had a special interest in the restoration efforts and a commitment to make them work. It was the abilities and enthusiasms of these people that led to the ultimate success of each of the mitigation projects. Without the commitment of the field engineer working for the Colorado Department of Transportation, the hydrologic control structure would not have been constructed as exactingly as it was, creating the appropriate hydrology for the sedge meadow that eventually developed. The university scientists and young wetland ecologists working for the Wisconsin Department of Transportation carefully followed and observed the construction process, helping make the necessary adjustments along the way which led to the success of that project. The Florida Department of Transportation's efforts to find new sites for mangrove plantings produced an unanticipated and spectacular success that was documented, years later, as a result of one participant's continuing interest in the project. At the Kachituli Oxbow, the construction managers were consultants hired by the project developer. The consulting staff had a long involvement in the project and the mitigation design. They were skilled scientists, committed to environmental protection and restoration, and they were supported by committed members of local environmental groups and representatives of the regulatory agencies. Combined, these human resources provided an essential ingredient to the success of wetland creation. The four case studies were successful partly because they modeled the natural prototypes very closely, but most importantly because they involved dedicated and knowledgeable individuals who ensured that during the planning, design, implementation, and management phases important ecological relationships were created and maintained.

The long-term success of the case studies and all the other mitigation projects implemented across the country will depend upon the future management of these restored landscapes. Each of the case study sites is owned and managed by public entities which, in some cases, will provide public access where appropriate and even modify them to better meet future needs of the watershed and the region. Each of the newly created landscapes was considered a success, yet what is more important is that the land has been set aside. The benefits of wetland functions can be garnered from these new landscapes to whatever extent is considered valuable. Whether these functions are optimized or not is less important than the preservation of these landscapes for future opportunities.

Many of the nation's wetlands have been, and more will continue to be, lost to the economic necessities of contemporary life. Each of the four projects compensated for wetlands lost to important and substantial development projects. The two highways and the bridge replacement project were reasonable and responsible solutions to problems created by population pressure, the effect rather than the cause of the inevitable population increases experienced everywhere in the country. The Florida Keys project, in particular, was conservative in design; rather than widening the bridges from two to four lanes, which would certainly have increased traffic into the fragile Keys environment, the project was restricted to a safer, more substantial duplication of the original two-lane bridges. The Colorado project, likewise, produced a wider, safer roadbed without increasing the capacity of the highway; and the beltway serving Madison, Wisconsin was way overdue in servicing its auto-dependent clients. In Yolo County, California, economic development was a reasonable goal that would have been served well by the imaginative planned development. Economic development cannot be stopped nor should it be. It can proceed in orderly and responsible fashion, however, as it did in the four case histories, providing a mechanism for solid wetland gains that can and often do extend beyond a simple compensation for wetland losses.

INDEX

A

B

I

K

L

M

N

P

R

S

Y